High Performance and Optimum Design of Structures and Materials VI

WIT*PRESS*

WIT Press publishes leading books in Science and Technology.
Visit our website for the current list of titles.
www.witpress.com

WIT*eLibrary*

Home of the Transactions of the Wessex Institute.
Papers contained in this volume are archived in the WIT eLibrary in volume 216 of WIT Transactions on the Built Environment (ISSN 1743-3509).
The WIT eLibrary provides the international scientific community with immediate and permanent access to individual papers presented at WIT conferences.
Visit the WIT eLibrary at www.witpress.com.

INTERNATIONAL CONFERENCE ON HIGH PERFORMANCE AND OPTIMUM DESIGN OF STRUCTURES AND MATERIALS

HPSM/OPTI 2025

Conference Chairmen

Santiago Hernández
University of A Coruña, Spain

Stavros Syngellakis
Wessex Institute, UK

International Scientific Advisory Committee

Organised by

Wessex Institute, UK

University of A Coruña, Spain

Sponsored by

WIT Transactions on the Built Environment

WIT Transactions

Wessex Institute
Ashurst Lodge, Ashurst
Southampton SO40 7AA, UK

We would like to express our thanks to the conference Chairs and members of the International Scientific Advisory Committees for their efforts during the 2025 conference season.

Conference Chairs

International Scientific Advisory Committee Members 2025

High Performance and Optimum Design of Structures and Materials VI

Editors

Santiago Hernández
University of A Coruña, Spain

Stavros Syngellakis
Wessex Institute, UK

WIT*PRESS* Southampton, Boston

Editors:

Santiago Hernández
University of A Coruña, Spain

Stavros Syngellakis
Wessex Institute, UK

Published by

WIT Press
Ashurst Lodge, Ashurst, Southampton, SO40 7AA, UK
Tel: 44 (0) 238 029 3223; Fax: 44 (0) 238 029 2853
E-Mail: witpress@witpress.com
http://www.witpress.com

For USA, Canada and Mexico

Computational Mechanics International Inc
25 Bridge Street, Billerica, MA 01821, USA
Tel: 978 667 5841; Fax: 978 667 7582
E-Mail: infousa@witpress.com
http://www.witpress.com

British Library Cataloguing-in-Publication Data

A Catalogue record for this book is available
from the British Library

ISBN: 978-1-78466-497-8
eISBN: 978-1-78466-498-5
ISSN: 1746-4498 (print)
ISSN: 1743-3509 (on-line)

The texts of the papers in this volume were set individually by the authors or under their supervision. Only minor corrections to the text may have been carried out by the publisher.

Preface

This volume contains a selection of the contributions presented at the International Conference on High Performance and Optimum Design of Structures and Materials (HPSM/OPTI) which took place from 10 to 12 June 2025. This edition of the scientific event follows the success of a number of meetings on structures and materials, and on optimum design, which originated in Southampton as long ago as 1989. As the meetings evolved, they gave rise to the current series, which started in Ostend (Belgium) in 2014 and continued in Siena (Italy) in 2016 and Ljubljana (Slovenia) in 2018. The health crisis created by the COVID-19, that reduced substantially people´s mobility, forced the transformation of the 2020 and 2022 conferences into online forums.

The 2025 edition of the conference was organised by the Wessex Institute and the University of A Coruña, Spain; the venue selected was Edinburgh, the beautiful capital city of Scotland.

This volume presents recent research initiatives on the development of materials with enhanced capabilities at micro, meso and macroscale, suitable for use in various scientific and technological applications. The focus is on characterisation and search for improved, mainly mechanical properties of different types of materials such as metals, plastics, composites and biomaterials. Bioengineering studies are also presented in the topic of molecular structural analysis.

Innovative structural typologies are also another area of interest in the book. There is more emphasis on sustainability rather than on innovations in large scale construction projects, such as high-rise buildings, large stadiums and sport halls, pedestrian bridges and long span bridges. Cardboard microarchitecture through scaled prototypes is proposed for internal space redevelopment leading to optimum life-cycles. Technological innovations contribute to meta-design of modular, energy-efficient, and flexible facilities for vulnerable populations' healthcare.

Both materials and structures can benefit from the advantages of optimisation techniques that guide the engineers and scientists in the search of the best solution for the problem they are working on. In that regard, numerical optimisation methods need to be seen as design tools that help engineering intuition, make better designers and speed up the learning curve of young practitioners. Currently these techniques are mature enough to be used in real design problems in mechanical, aerospace, construction or civil engineering. Several commercial codes have a variety of optimisation capabilities for various engineering fields. The application of various optimisation strategies is presented in this volume: structural, shape and topology and multi-objective optimisation as well as evolutionary methods.

The contributing authors are experts from many countries specialising in fields of materials definition and characterisation, structural designers, architects, mechanical, aircraft and civil engineers as well as experimentalists and computer analysts developing software for application to the fields covered in the book.

As with previous Wessex Institute HPSM/OPTI conference proceedings, this volume is part of the WIT Transactions on the Built Environment series and its papers are archived online in the WIT elibrary (www.witpress.com/elibrary), where they are freely available to the international scientific community.

It is expected this volume to be a valuable addition to the existing literature on the subject and will be of interest to any researcher working in this scientific field.

The editors are grateful to the authors for their contributions and to the members of the International Scientific Advisory Committee of the conference for reviewing the submitted abstracts and manuscripts thus ensuring the high quality of the contents of this book.

The Editors, 2025

Contents

Section 9: Structural optimisation

SECTION 1
COMPOSITE MATERIALS

RESIN IMPREGNATION FOR CONTINUOUS CARBON FIBRES USING ROLLERS

YUJI KONDO, KOYA HOSOKAWA, TATSUYA TANAKA & MASAHIRO SASADA
Department of Mechanical Engineering, Graduate School of Science and Engineering,
Doshisha University, Japan

ABSTRACT

In the pursuit of high-quality and high-strength continuous fibre-reinforced thermoplastic (CFRTP) filaments, enhancing fibre dispersibility is critical. This study focuses on enhancing the fibre spreading mechanism via the roller method, a process that was previously inaccessible to observation due to the use of resin-filled equipment. To overcome this limitation, an experimental apparatus was developed that enabled in situ visualization of fibres' behaviour. In the conventional roller spreading method, the cylindrical rollers had limitations in fibre spreading. An alternative approach employing inclined rollers was then tested; however, excessive inclination led to a pronounced central bias of the fibres, further hindering uniform dispersion. To address this challenge, we devised a roller with a curved shape designed to mitigate fibre bias while increasing the spreading width. Continuous fibre-reinforced filaments were produced using this innovative geometry, and the effects on fibre dispersibility and void area were evaluated at pull speeds of 1 m/min and 2 m/min. The findings of the study demonstrated that the curved roller reduced the void area and a modest enhancement in fibre dispersibility when compared to traditional cylindrical rollers. These observations suggest that the curved roller design is effective in producing superior quality CFRTP filaments and indicate a promising potential for further improvements in composite material performance.

Keywords: CFRTP, fibre spreading, roller spreading method.

1 INTRODUCTION

Carbon fibre reinforced thermoplastics (CFRTPs) have attracted attention as a promising material due to their high mechanical properties, short moulding cycle times, and recyclability. These characteristics render CFRTPs well-suited for the reduction of weight in large transportation equipment, such as automobiles, by replacing metal components and thereby reducing environmental impact [1].

One of the manufacturing methods for CFRTP is pultrusion, wherein continuous carbon fibre bundles are drawn through a resin-impregnation bath containing molten thermoplastic resin. This process impregnates the fibre bundles with resin, forming CFRTP strands that serve as intermediate materials, such as long-fibre pellets or unidirectional (UD) tapes, for various applications [2]. In the context of continuous-fibre CFRTP, it has been established that fibre spreading enhances fibre dispersion, thereby promoting impregnation and leading to enhanced mechanical strength [3]. To achieve fibre spreading, roller-based or air-jet methods are commonly employed prior to the impregnation process [4].

In this study, the focus is on a molten-impregnation method that utilizes a roller-based approach (henceforth referred to as the 'roller-spreading method') to enable simultaneous fibre spreading and resin impregnation. Specifically, a spreading roller is placed inside the molten resin bath, allowing both fibre spreading and resin impregnation to occur simultaneously. The underlying mechanism of roller spreading is as follows: when a fibre bundle, impregnated with resin, is pulled out from the bath, a tensile force is generated. At the roller, this tension induces a compressive force on the fibre bundle, with the horizontal component of this compressive force acting on each filament, causing the bundle to spread. As this method relies on compressive force rather than the removal of sizing agents, bundles

WIT Transactions on The Built Environment, Vol 216, © 2025 WIT Press
www.witpress.com, ISSN 1743-3509 (on-line)
doi:10.2495/HPSM250011

with sizing can also be effectively spread. Furthermore, simultaneous execution of spreading and impregnation simultaneously enhances process productivity.

To investigate this roller-spreading method in detail, an experimental apparatus was constructed using silicone oil in place of molten thermoplastic resin, allowing for direct visualization of the spreading process. The behaviour of the fibre bundles during the spreading process was observed and the width of the bundles after spreading was measured. While conventional roller-spreading methods typically employ cylindrical rollers, previous studies have reported certain limitations associated with this roller geometry. Therefore, this study examined various roller shapes and the results are presented in the following sections.

2 ROLLER SHAPE TO IMPROVE FIBRE SPREADING WIDTH

2.1 Purpose of the experiment

The effects of impregnation bath parameters on fibre spreading using cylindrical rollers have been investigated. The experimental results with cylindrical rollers suggest that both the compressive force and the resin film influence the spread width. Therefore, to optimize fibre spreading, the focus was placed on roller geometry and the examination of shapes that could more effectively utilize compressive force. Specifically, the study explores roller geometries capable of achieving greater fibre spreading than cylindrical rollers. To evaluate the effectiveness of these rollers, they were fabricated using a 3D printer and the influence of roller shape on spread width was analysed.

2.2 Experimental methods

2.2.1 Experimental conditions

In this experiment, PAN-based carbon fibre bundles (TR50S15L, manufactured by Mitsubishi Chemical) were used. These fibre bundles consist of 15,000 filaments and have a width ranging from 6.0 to 8.0 mm. To visualize the interior of the impregnation bath, silicone oil (manufactured by Shin-Etsu Chemical Co., Ltd.) was used as a substitute for high-temperature molten resin. The silicone oil used in this experiment had a viscosity of 100,000 mm^2/s. The utilization of silicone oil is advantageous due to its colourless transparency and its relatively stable viscosity over a wide temperature range.

2.2.2 Fibre spreading experimental apparatus

The fibre spreading mechanism of the roller method remains unclear due to the inability to observe the interior of the resin impregnation bath. To address this issue, a transparent mock experimental device with the same structure as the molten resin impregnation bath was constructed (hereafter referred to as the resin impregnation bath). The structure of the resin impregnation bath is shown in Fig. 1.

In the experimental setup, a carbon fibre bundle is drawn from the carbon fibre roving, passed through the resin impregnation bath, and then wound up by a take-up motor. A high-resolution camera (Canon EOS M10) was positioned above the resin impregnation bath to capture the spreading behaviour. Initial tension (rear tension) was applied using a torque motor, as illustrated in Fig. 2.

The rollers were fabricated from ABS resin with a surface roughness of Ra 0.05. For the take-up motor, a brushless motor (BLE2D120-A, Oriental Motor Co.) was employed. Each experiment was conducted for a total duration of 6 seconds (recording time), and for all experiments described below, the number of trials was set to $N = 5$.

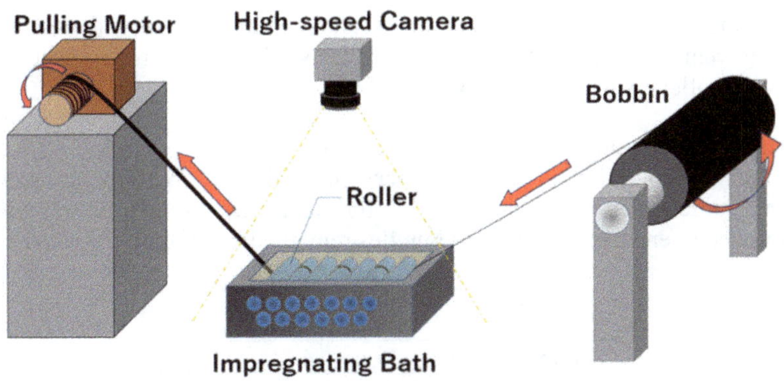

Figure 1: Model of impregnation bath.

Figure 2: Equipment for back tension.

2.2.3 Methods for assessing fibre bundle spreading width

The experiment was conducted by operating the motor for a period of 6 seconds before stopping. The 6-second duration was selected to prevent alterations in fibre spreading behaviour due to the depletion of silicone oil. Subsequent to each experiment, images captured by a high-resolution camera were analysed using the software SIGMASCAN PRO to evaluate the spreading width of the fibre bundles. Due to the structural constraints of the resin impregnation bath, measurements were limited to the spreading width at every even-numbered roller.

2.2.4 3D printing of rollers

An FDM-type 3D printer (S3DP 222, S-Lab) was used to produce the rollers. The printing conditions are summarized in Table 1. ABS filament (diameter: 1.75 mm) was used as the roller material. The printing temperature was set at 260°C, and the bed temperature was maintained at 110°C. Nozzle moving speed was 2 mm/s.

Table 1: 3D print condition.

Nozzle temperature (°C)	260
Bed temperature (°C)	110
Nozzle moving speed (mm/s)	2
Layer height (mm)	1.1

After printing, the rollers were placed in a sealed container filled with acetone vapor for 40 minutes to remove layer marks from the roller surface. The surface roughness of the acetone-treated rollers was measured to be Ra 0.05.

2.2.5 Experimental conditions and roller shapes

The experimental conditions are summarized in Table 2. With regard to the roller arrangement, guide rollers were positioned before and after the test rollers, with different shapes employed to ensure that the fibre bundles remained centred on the roller (see Fig. 3).

Table 2: Experiment conditions.

Number of rollers	4
Pulling speed (m/min)	3.7
Roller diameter (mm)	13
Back tension (N)	5
Roller condition	Rotation

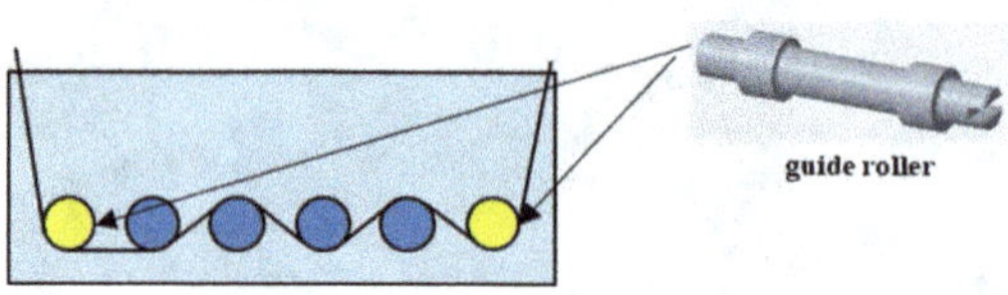

Figure 3: Rollers position.

The hypothesis was that the force acting in the spreading direction on the fibre filaments would increase by introducing an inclination to the roller, thereby enhancing fibre spreading. To investigate this hypothesis, two types of rollers were fabricated: one with a linear longitudinal profile and another with a circumferential longitudinal profile defined by an arc of circumference. The schematics and dimensions of the fabricated rollers are illustrated in Figs 4 and 5, respectively.

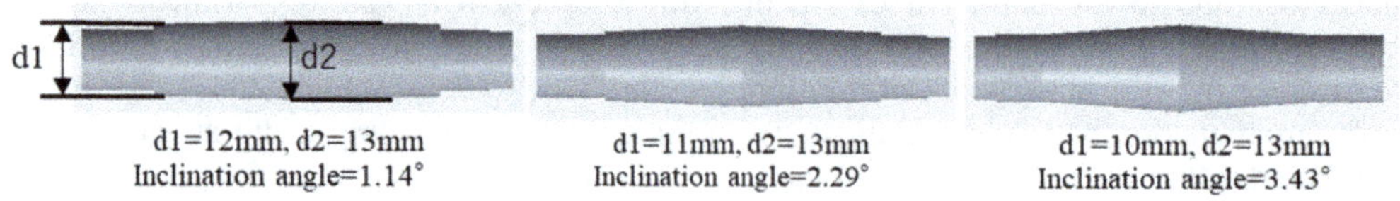

Figure 4: Roller with linear longitudinal profile.

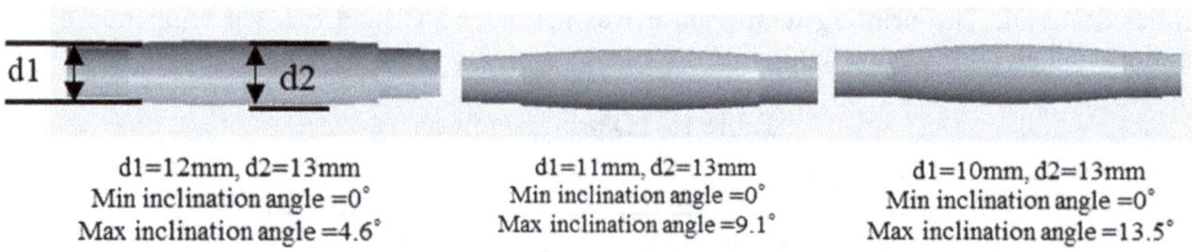

Figure 5: Roller with circumferential longitudinal profile.

The roller with a linear longitudinal profile, the purpose of which was to study the influence of the inclination angle, was designed with the following specifications: the maximum diameter of 13 mm at the central section, and three inclination angles were tested (1.14°, 2.29° and 3.43°). The roller with a circumferential profile was designed to examine the effect of the slope shape on fibre spreading. The maximum diameter at the central section of this roller was set to 13 mm, while the minimum diameters at both ends were set to 10 mm, 11 mm and 12 mm, respectively. This configuration resulted in a continuous circumferential profile connecting the centre and the ends of the roller.

2.3 Results and discussion

2.3.1 Experimental results for rollers with linear profile

The experimental measurements for fibre spreading width and tensile force obtained using rollers with a linear profile (hereafter referred to as 'linear rollers') are presented in Fig. 6. Fibre misalignment was observed during the experiment, as illustrated in Fig. 7. Therefore, both the spreading width measured with the misalignment and the spreading width measured without the misalignment were recorded.

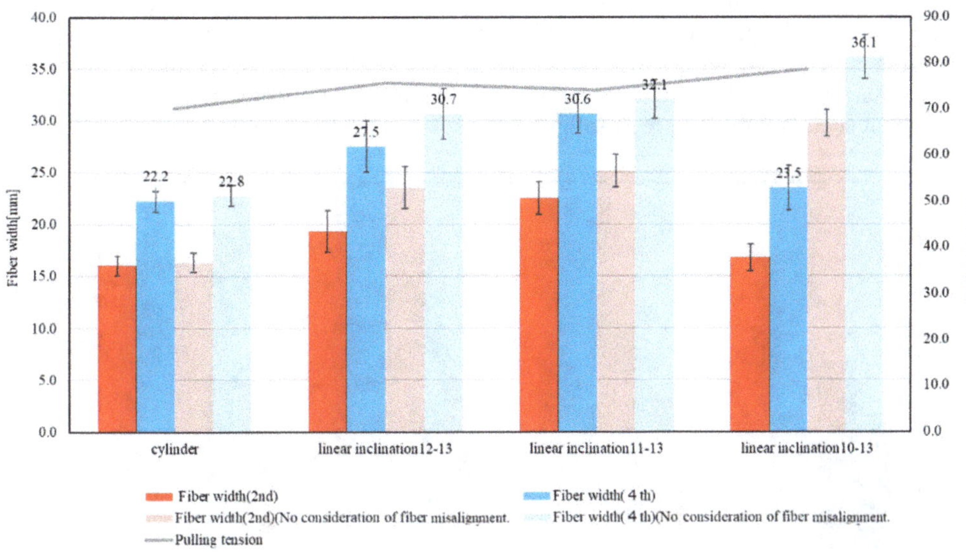

Figure 6: Experiment results using roller with linear profile (average and standard deviation of measurements).

For a slope angle of 1.14°, the average spreading width at the fourth roller was measured as 27.5 mm when considering misalignment and 30.7 mm when not considering it, with a tensile force of 75.8 N. For a slope angle of 2.29°, the average spreading width at the fourth roller was 30.6 mm considering misalignment and 32.1 mm not considering it, with a tensile force of 74.2 N. For a slope angle of 3.43°, the average spreading width at the fourth roller was 23.5 mm considering misalignment and 36.1 mm not considering it, with a tensile force of 78.7 N. In all cases, the spreading width exceeded that achieved using cylindrical rollers.

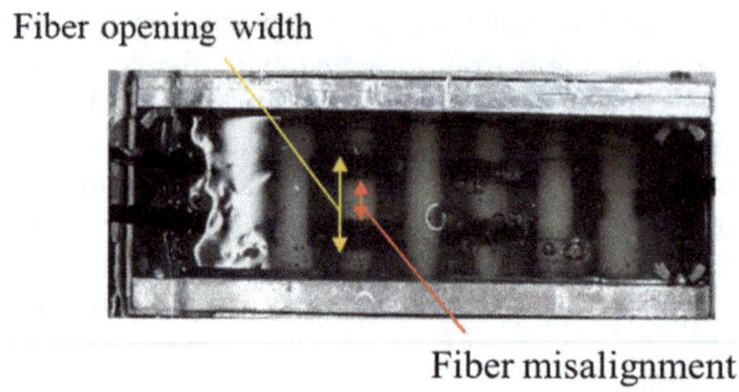

Figure 7: Experiment for rollers with linear profile.

Furthermore, the experiments confirmed that fibre misalignment became more pronounced as the slope angle increased. Fig. 7 also indicates that this misalignment occurs primarily around the central region of the roller.

2.3.2 Experimental results for rollers with circumferential longitudinal profile

Fig. 8 presents the experimental results for fibre spreading width and tensile force obtained using rollers with circumferential longitudinal profile (hereafter referred to as ‘curved rollers’). As with the rollers featuring a linear profile, the spreading width was measured with and without accounting for fibre misalignment.

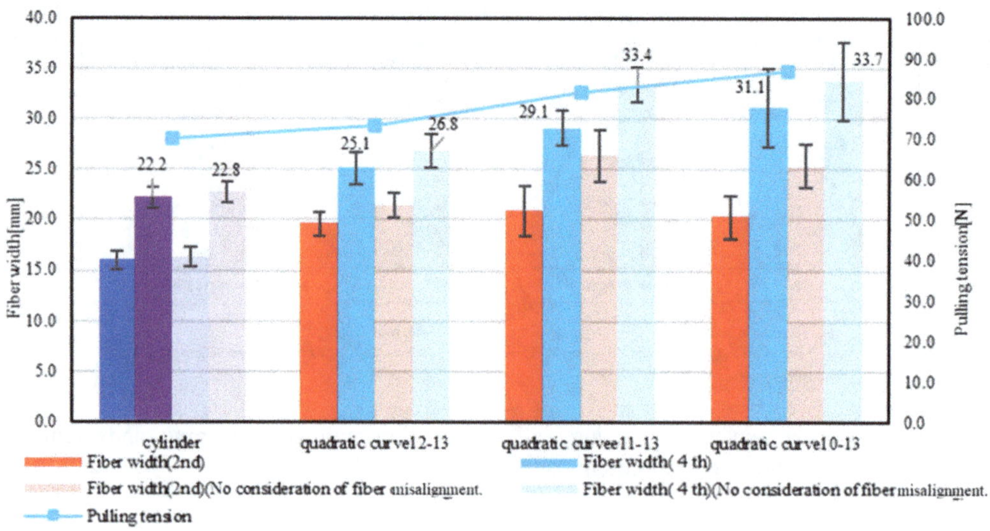

Figure 8: Experiment results using rollers with circumferential profile (average and standard deviation of measurements).

For the profile 12–13 (maximum slope angle of 4.6°, see Fig. 5), the average spreading width at the fourth roller was measured to be 25.1 mm when considering misalignment and 26.8 mm when not considering it, with a tensile force of 73.5 N. With a maximum slope angle of 9.1° (profile 11–13), the average spreading width was 29.1 mm considering misalignment and 33.4 mm not considering it, with a tensile force of 81.5 N. The profile 10–13 (maximum slope angle of 13.5°) showed an average spreading width of 31.1 mm considering misalignment and of 33.7 mm not considering it, with a tensile force of 86.8 N.

As illustrated in Fig. 9, the experimental image reveals that there was barely visible occurrence of fibre misalignment, thereby demonstrating that the fibres were distributed in an ideal manner for impregnation. Even at the maximum slope angle, no misalignment was observed, suggesting that the curved geometry effectively suppresses fibre misalignment. Furthermore, across all tested dimensions, the curved rollers achieved a greater spreading width than the cylindrical rollers.

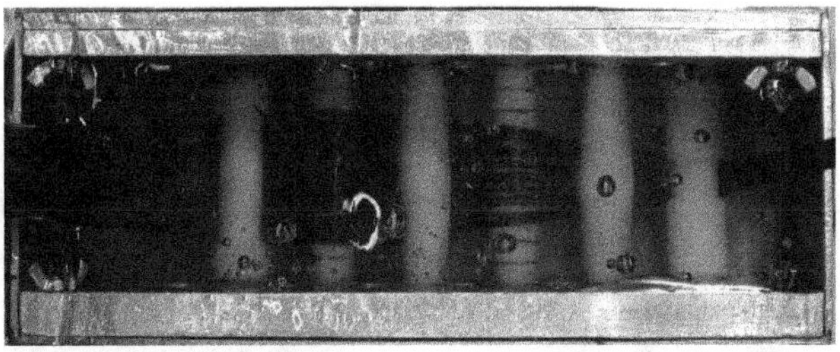

Figure 9: Experiment for rollers with circumferential profile.

3 EVALUATION USING 3K CARBON FIBRE BUNDLES

The aim of this phase of the research was to fabricate and evaluate samples using a roller with circumferential profile, which was found to enhance fibre spreading width. 3K carbon fibre bundles were used for manufacturing. By assessing the quality of the filaments produced by the actual machine, the effectiveness of the curved roller shape was validated.

3.1 Experimental materials, setup and measurements

Polyetherimide (PEI) was used as the matrix material, while PAN-based carbon fibre tow, consisting of 3,000 filaments, was used as the reinforcement material. The diameter of the carbon fibres was 7 µm.

The experimental conditions are outlined in Table 3. Five rollers were placed in the impregnation bath, with the second, third, and fourth rollers used to compare cylindrical and curved rollers.

Table 3: Experiment conditions.

Resin	PEI
Number of fibres	3K
Resin temperature (°C)	400
Number of pins	5
Speed (m/min)	1, 2

The cross-sections of the fabricated filaments were observed using imaging analysis. A compact scanning electron microscope (SEM) (TM3030+) was employed for the observations. Prior to imaging, the samples were embedded in resin and polished to obtain a smooth cross-sectional surface.

To analyse the void area, the image processing software Sigma Scan Pro 5 was used to measure the number of pixels corresponding to void regions, from which the void area was calculated. A smaller void area indicates higher impregnation quality.

To evaluate fibre dispersion, the centroid distance of carbon fibres (Voronoi region) was utilized, as shown in Fig. 10. A Voronoi diagram partitions a plane based on the proximity of each point to its nearest neighbouring point. The homogeneity coefficient H was calculated using the average area and standard deviation of the Voronoi regions and is expressed as follows [5]:

$$H = \frac{\text{Standard deviation}}{\text{Average area}} \tag{1}$$

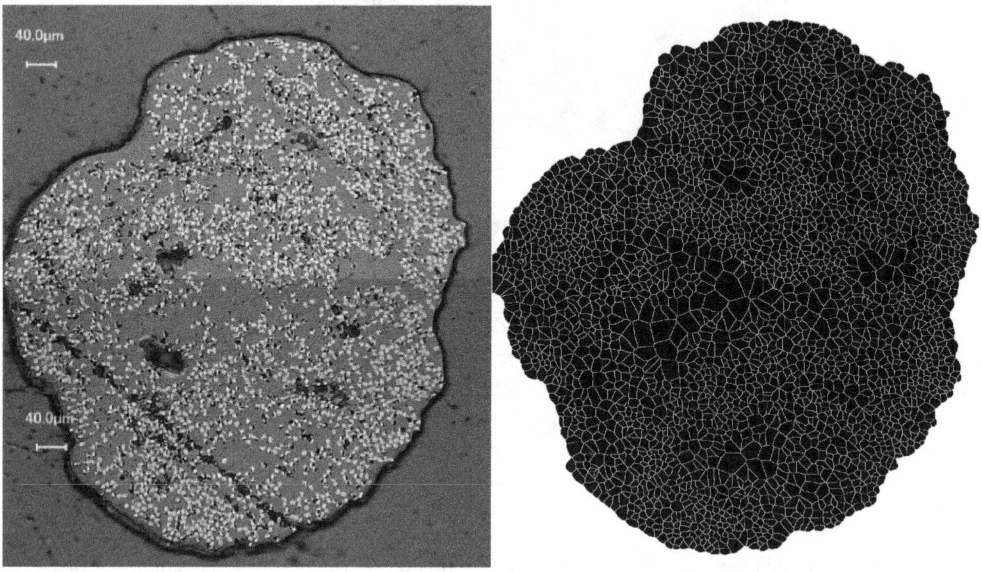

(a) (b)

Figure 10: SEM observations. (a) Cross-sectional image; and (b) Voronoi diagram.

A smaller homogeneity coefficient *H* indicates better fibre dispersion. The number of measured samples was six.

3.2 Experimental results

3.2.1 Measurement of void area

The results of the void area measurements obtained from the cross-sectional observations of the filaments are presented in Figs 11 and 12. Fig. 11 compares the void areas of filaments produced using cylindrical and curved rollers at a take-up speed of 1 m/min, while Fig. 12 presents the comparison at a take-up speed of 2 m/min.

At a take-up speed of 1 m/min, the average void area of the filaments was 0.000132 mm^2 for the cylindrical roller and 0.000071 mm^2 for the curved roller. Similarly, at 2 m/min, the average void area was 0.000244 mm^2 for the cylindrical roller and 0.000131 mm^2 for the curved roller.

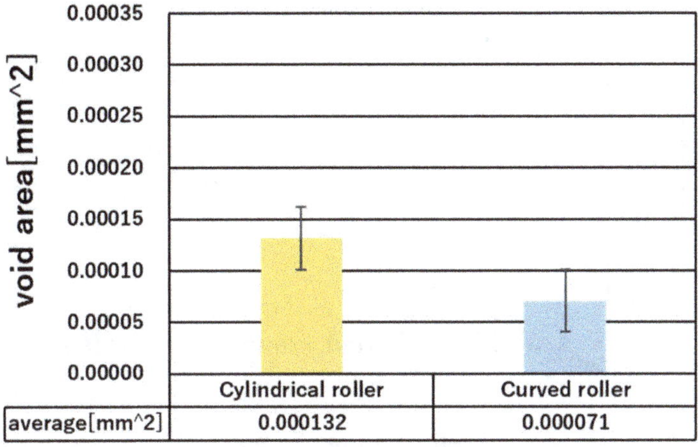

Figure 11: Void area at 1 m/min (average and standard deviation of six measurements).

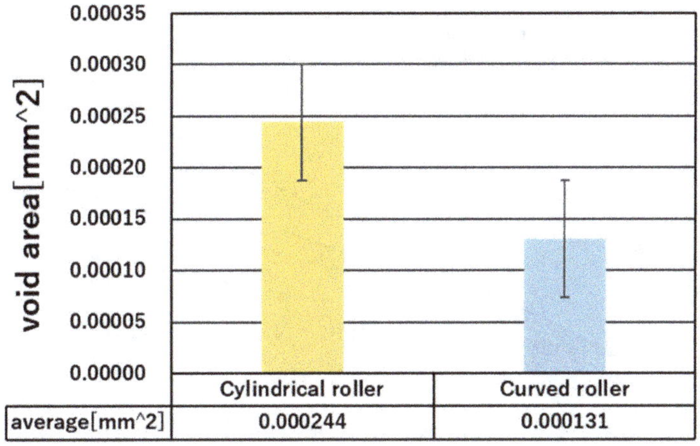

Figure 12: Void area at 2 m/min (average and standard deviation of six measurements).

An increasing void area was observed as the take-up speed increased. Furthermore, for both take-up speeds, the utilization of the curved roller resulted in a reduction in void area. Notably, even at a take-up speed of 2 m/min, the void area of filaments produced using the curved roller was reduced to a level comparable to that of filaments produced at 1 m/min using the cylindrical roller.

3.2.2 Fibre dispersibility

The homogeneity coefficients of fibre dispersibility in the filament cross-sections at take-up speeds of 1 m/min and 2 m/min are presented in Figs 13 and 14, respectively.

At a take-up speed of 1 m/min, the homogeneity coefficient was 0.83 for the cylindrical roller and 0.75 for the curved roller. A similar trend was observed at 2 m/min, the homogeneity coefficient was 0.75 for the cylindrical roller and 0.69 for the curved roller.

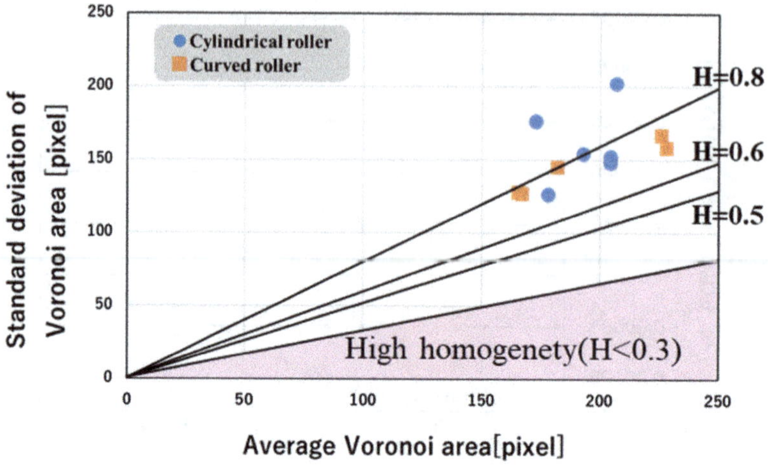

Figure 13: Results of fibre dispersibility homogeneity: take over speed 1 m/min.

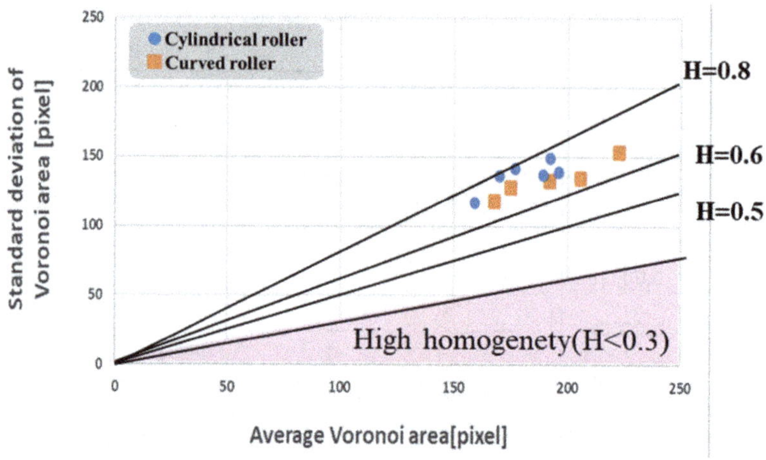

Figure 14: Results of fibre dispersibility homogeneity: take over speed 2 m/min.

Despite the use of the curved roller resulted in a reduction in the homogeneity coefficient, no substantial enhancement in fibre dispersion was observed.

4 CONCLUSIONS

The aim of this study was to improve fibre dispersibility in the production of high-quality, high-strength CFRTP filaments. In contrast to conventional resin-impregnation methods, experiments were conducted using an apparatus that allowed in situ visualization of fibre behaviour.

Initial experiments with conventional cylindrical rollers confirmed their limitations in fibre spreading, while trials with linear and circumferential longitudinal profile rollers revealed that excessive inclination resulted in a pronounced central bias of fibres. In response, a novel roller with a curved slope was devised with the aim of mitigating fibre bias and expanding the spreading width. Testing at pull speeds of 1 m/min and 2 m/min demonstrated that the curved roller significantly reduced the void area and provided a modest improvement in fibre dispersibility compared to traditional cylindrical rollers.

The findings indicate that the curved roller design offers a promising approach to enhancing CFRTP filament quality.

REFERENCES

[1] Terada, K., Carbon fibre reinforced thermoplastics (CFRTP). *Journal of the Japan Society for Precision Engineering*, **81**(6), 2015.

[2] Wakazono, T., Impregnation equipment for thermoplastic plastics. *Kobe Steel Industry Reports*, **67**(2), 2018.

[3] Tanaka, T., Fibre lengthening and dispersion technology in fibre-dispersed thermoplastic composite materials. *Journal of the Japan Society of Colour Material*, **86**(6), pp. 231–236, 2013.

[4] Kawabe, K. & Matsuo, T., Fibre bundle spreading techniques and applications to composite materials. *Journal of the Japan Society for Composite Materials*, 51(2), 2018.

[5] Hopmann, C., Wilms, E., Beste, C., Schneider, D., Fischer, K. & Stender, S., Investigation of the influence of melt-impregnation parameters on the morphology of thermoplastic UD-tapes and a method for quantifying the same. *Journal of Thermoplastic Composite Materials*, **34**(9), pp. 1299–1312, 2019.

ELONGATIONAL FLOW EFFECT ON THE PERFORMANCE OF BIOMASS POLYMER ALLOYS

RYO SEIRIKI, BIREI TANAKA, TATSUYA TANAKA & MASAHIRO SASADA
Department of Mechanical Engineering and Science, Doshisha University, Japan

ABSTRACT

In the context of environmental conservation, there is an ongoing transition from petroleum-derived plastics to biomass-derived plastics, a shift that encompasses the field of engineering plastics. One example is bio-polycarbonate (Bio-PC). The aim of this study was to enhance the impact strength of PMMA, an acrylic resin that is characterized by brittleness, by alloying it with Bio-PC, a ductile material with excellent impact resistance. In the PMMA/Bio-PC system, which is incompatible and has a two-phase morphology, the dispersion and distribution of the dispersed phase are important for the development of impact resistance properties. In the PMMA/Bio-PC system with different viscosity ratios, the kneading method employing elongational flow has been shown to be effective when composites are produced using a twin-screw extruder. Therefore, a comparison was made between conventional shear kneading and elongational kneading. The results showed that the kneading method using elongational flow was effective in terms of dispersibility, although no significant difference in distributivity was observed. It was found that there is a possibility that further improvement of distributivity in the kneading method using elongational flow can lead to the development of enhanced impact properties.

Keywords: polymer alloy, elongational flow, twin screw extruder, impact strength.

1 INTRODUCTION

A transition from petroleum-derived plastics to biomass-derived plastics is occurring in the context of environmental conservation [1]. This transition encompasses the field of engineering plastics, with notable examples such as the development of Bio-polycarbonate (Bio-PC). Bio-PC is a ductile material with enhanced impact resistance. The alloying of Bio-PC with PMMA, an acrylic resin with inferior impact strength, allows for the creation of a material with complementary properties. The present study aims to enhance the impact strength of PMMA and Bio-PC composites. The incompatibility between PMMA and Bio-PC results in the formation of a Bio-PC phase, which does not dissolve into the PMMA. In the PMMA/Bio-PC system, where a two-phase morphology is observed, the interface between the Bio-PC dispersed particles and the PMMA matrix, a region of concentrated stress when a load is applied, forms a shear zone by shear-yielding of the matrix [2]. This process may absorb the impact of energy, a phenomenon believed to contribute to impact strength [3]. A potential approach to enhance impact strength is to improve dispersibility by refining the dispersed phase [4]. The number of dispersed particles is known to increase with decreasing size of the dispersed phase, and it is expected that this would lead to an increase in shear zones, which would in turn increase the energy absorption rate and improve impact strength.

However, it should be noted that a barrier to impact strength exists, namely cracking between the dispersed particles. It is hypothesized that a critical particle-to-particle distance exists at which cracking occurs between dispersed particles, and that exceeding this distance leads to cracking between particles, thus impeding enhanced impact strength [5], [6]. To prevent fracture due to cracking, uniform distribution of the dispersed particles is necessary [5].

www.witpress.com, ISSN 1743-3509 (on-line)
doi:10.2495/HPSM250021

In polymer composites, the process of melt compounding using a twin-screw extruder is frequently employed. The flow patterns within the twin-screw extruder are characterized by the existence of shear flow and elongational flow, with shear flow exhibiting a dominant presence [7]. The dispersion of PMMA/Bio-PC systems with different viscosities, through the utilization of shearing, has proven to be a challenge [8]. Conversely, it has been observed that elongational flow possesses the capacity to disperse PMMA/Bio-PC systems with different viscosity ratios [8]. The formation of a fine microstructure through elongational flow is expected to stabilize the flow by homogenizing the viscosity and improve the distribution of the dispersed phase. To create an elongational flow field in a twin-screw extruder, a special segment fixed blister disk (XBD) [9] was developed by the authors. After the dispersed phase is dispersed using XBD, the dispersed phase is evenly distributed using a full-flight screw with excellent transport and distribution properties.

In this study, a comparison was conducted between the dispersibility and distributivity of PMMA/Bio-PC systems by elongational kneading using XBD and conventional kneading using kneading disk (KD). Subsequently, to investigate the effect of elongational mixing on the impact strength of the PMMA/Bio-PC system, samples were prepared via elongational mixing and subjected to impact testing.

2 PRELIMINARY KNOWLEDGE

2.1 Comparative experiments between shear kneading and elongational kneading

2.1.1 Theory

Elongational flow is a flow in which a velocity distribution is generated in the flow direction. The dispersion effect of elongational flow in blends of incompatible Newtonian fluids has been demonstrated by Grace [10]. In this experiment, the shear and elongational splitting conditions of a single droplet in a matrix liquid were obtained experimentally, as shown in Fig. 1. The vertical and horizontal axes are the capillary number Ca and the viscosity ratio ρ, respectively, expressed by eqn (1):

$$Ca = {}^{\eta_M \dot{\gamma} \rho}/_{\sigma} \quad , \quad \rho = {}^{\eta_D}/_{\eta_M}. \tag{1}$$

Here, η_M is the matrix viscosity, η_D is the dispersed phase viscosity, $\dot{\gamma}$ is the strain rate and σ is the interfacial tension. Ca is the ratio of the force generated by the matrix flow to the interfacial tension. In other words, it can be recognized as the force that tries to deform the droplet by flow and the force that tries to undo the deformation by the interfacial tension. The solid line in Fig. 1 shows the critical capillary number at which the droplet splits into simple shear flow and simple elongational flow, and the droplet can disperse in the region above the solid line. The results show that the elongational flow is below the critical Ca of the shear flow for all viscosity ratios, which means that the elongational flow can disperse in the region above the solid line. This suggests that elongational flow may be more energetically efficient than shear flow in terms of dispersion. In addition, shear flow cannot split at viscosity ratios above about 5, no matter how much shear action is applied, whereas elongational flow can split over a wide range of viscosity ratios. For this reason, elongational flow has attracted attention in the field of composites. However, shear flow is dominant in melt kneading using a twin-screw extruder, which is often used for composites, and it is necessary to intentionally

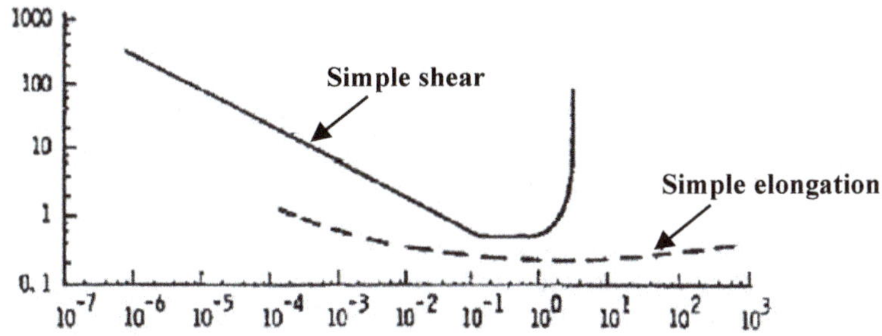

Figure 1: Relation between critical Capillary number and viscosity ratio according to Grace's analysis [10].

generate elongational flow [7]. In previous studies, the rapid expansion and contraction geometry has been used to obtain high elongation rates as a kneading technique using elongational flow [11]. If the polymer flows through a channel with compressive and expansive sections, the flow changes abruptly in the compressive and expansive sections, resulting in a large change in the elongation velocity. As shown in eqn (2) [7], an increase in the elongational strain rate leads to higher elongational stress, which is expected to promote dispersion. In eqn (2), σ_E represents elongational stress, η_E denotes the elongational viscosity, and $\dot{\varepsilon}$ is the elongational strain rate.

$$\sigma_E = \eta_E \dot{\varepsilon}. \tag{2}$$

In this study, the XBD segment (see Fig. 2) was used to generate the predominant elongated flow field within the twin-screw extruder. The XBD has multiple through holes with a diameter of 1.0 mm or less. As shown in Fig. 2, upon insertion of the XBD between the Full Flight Screw, which is a conveying screw, the resin conveyed flows into the multiple through holes. Consequently, the elongation velocity increases with the loss of pressure due to the rapid shrinkage of the flow path, thereby creating a dominant elongational flow field. In this study, twin-screw extrusion experiments were conducted using a screw configuration with an XBD and a conventional KD. These experiments were undertaken to demonstrate the effectiveness of elongational flow. Initially, PMMA/Bio-PC was kneaded using the Full Flight segment screw configuration, which has been shown to have reduced kneading action and superior distribution and conveying capabilities, to prepare materials for comparison.

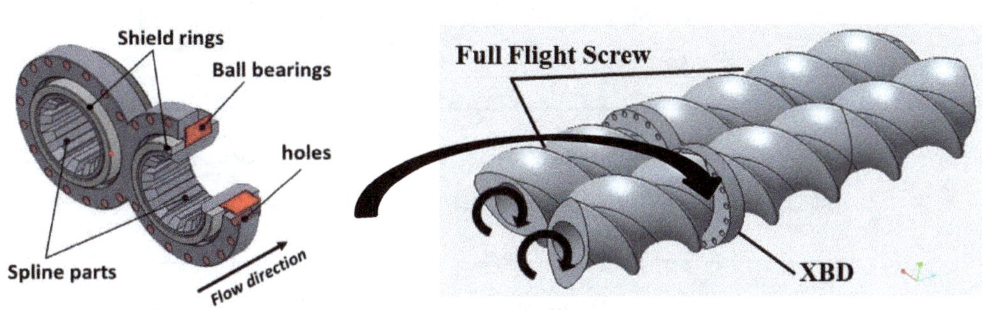

Figure 2: Fixed blister disk (XBD) [9].

Subsequently, the screw configuration was used for elongational kneading with XBD and for shear kneading with KD.

3 MATERIALS AND EXPERIMENTAL PROCEDURES

3.1 Shear kneading and extension kneading

The experimental apparatus comprised a twin-screw extruder (ZSK18, manufactured by Cooperion, Inc.), with a screw diameter of 18 mm, L/D = 40, and a maximum torque of 38 Nm. The operating conditions are detailed in Table 1. The material composition ratio was 50/50 (wt%). Polymethyl methacrylate resin: PMMA (ACRYPET1, Mitsubishi Chemical Corp., VH001) and Bio-polycarbonate: Bio-PC (DURABIO, Mitsubishi Chemical Corp., D7340R) were used as materials. Bio-PC has a density of 1.36 g/cm^3 and an MFR of 2.0 g/10 min (230°C, 3.8 kg). In the experiments, this material was previously kneaded using the full-flight screw configuration shown in Fig. 3(I), which exhibited minimal kneading action. Using the obtained samples, shearing and elongation mixing were carried out. The screw configuration for elongational kneading with XBD is shown in Fig. 3(II), while the screw configuration for shearing kneading with KD is shown in Fig. 3(III). After the extrusion experiments, morphological observations were conducted to compare dispersibility and distribution. The extruded pellets were moulded into a film (100 μm) by a press machine. After moulding, the morphological structure was observed using an X-ray CT (nano3DX, Rigaku Corporation).

Table 1: Operating conditions.

Composition ratio (wt%)		Quantity:	Number of rotations:	Barrel temperature
PMMA	Bio-PC	Q(kg/h)	Ns (rpm)	(°C)
50	50	2.5	150	220

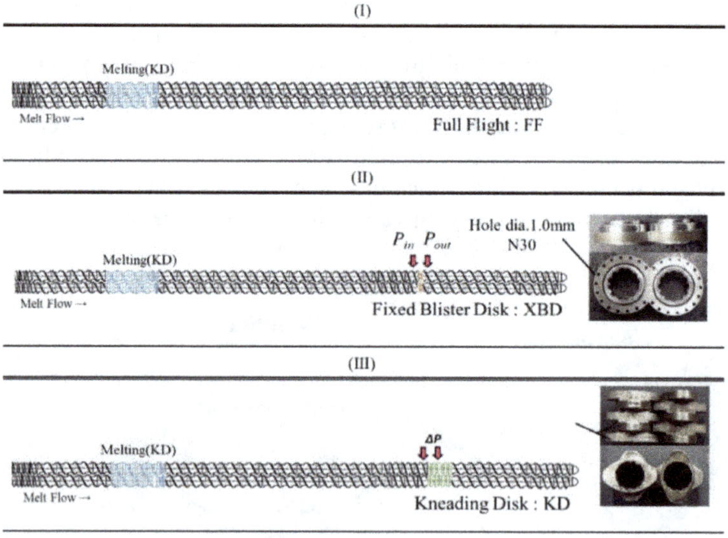

Figure 3: Screw configuration. (I) Distributive extrusion; (II) Elongational extrusion; and (III) Shear extrusion.

3.2 Evaluation of Impact strength

Samples of PMMA/Bio-PC with different composition ratios were prepared by elongational kneading, and dumbbell specimens were moulded using a small injection moulding machine. The specimens were then subjected to Izod testing. A co-axial twin-screw extruder (Thermo Fisher Scientific TSE24 MC Twin-screw Extruder Barrel Dia24, L/D:28) was used for kneading. The screw configuration shown in Fig. 4 was used in the design of the extruder to achieve adequate impact strength. The composition ratio of PMMA/Bio-PC was 90/10 wt%, 80/20 wt%, and 60/40 wt%. The kneading temperature was set at 220°C, the feed rate at 3 kg/h, and the rotation speed at 125 rpm. The pellets obtained from twin-screw extrusion were fed into an injection moulding machine (Thermo Fisher Scientific HAAKE Mini Jet Pro) to produce dumbbell specimens. An Izod impact tester (Impact Tester IT, Toyo Seiki Co., Ltd.) was used for impact resistance tests.

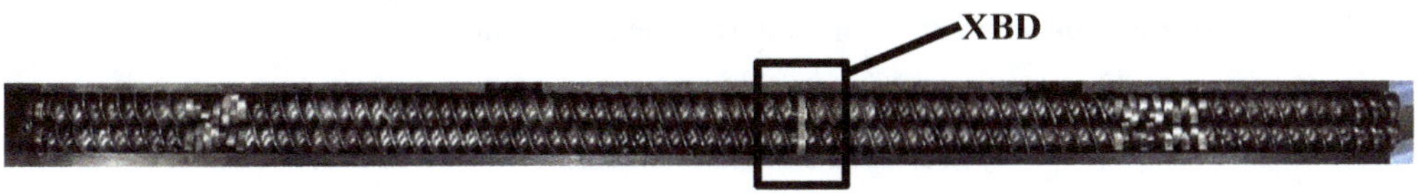

Figure 4: Screw configuration.

4 RESULTS AND DISCUSSION

4.1 Comparison of dispersibility and distributivity between shear kneading and elongational kneading

The morphology of the samples obtained by shear kneading using conventional KD and by elongational kneading using XBD is shown in Fig. 5. Bio-PC with high relative density is white, while PMMA with low density is black. The material produces with Full Flight Screws without kneading does not form a sea-island structure and has a co-continuous structure (Fig. 5(a)), indicating poor dispersion. In contrast, the XBD sample in Fig. 5(b) and the KD sample in Fig. 5(c) form a sea-island structure, indicating that the XBD and KD segments have dispersion performance. Comparing the XBD and KD samples, the XBD sample tends to have a finer dispersed phase size than the KD sample. Fig. 6 shows the average dispersed phase size and frequency distribution of the XBD and KD samples. The average dispersive phase size of the XBD sample was 5.35 μm^2, and that of the KD sample was 39.9 μm^2. Fig. 6(a) shows that the dispersed phase of XBD is approximately 86% smaller than the KD sample. Furthermore, Fig. 6(b) and (c) show that the slope of the cumulative frequency distribution of the XBD sample is greater than that of the KD sample, indicating that the uniformity of the dispersed phase size is superior. These results indicate that elongational kneading using XBD is superior to shear kneading using KD in terms of making the size of the dispersed phase finer and more uniform. Moreover, the results suggest that elongational kneading using XBD is effective for the dispersibility of PMMA/Bio-PC composites.

A box-counting method using ImageJ software was used to obtain the fractal dimension and estimate distributivity. The observed image is equally divided into squares of side r, and a set of boxes is generated. If $N(r)$ is the number of boxes containing even a small amount of dispersive phase, it is known that the relation in eqn (3) holds [12]:

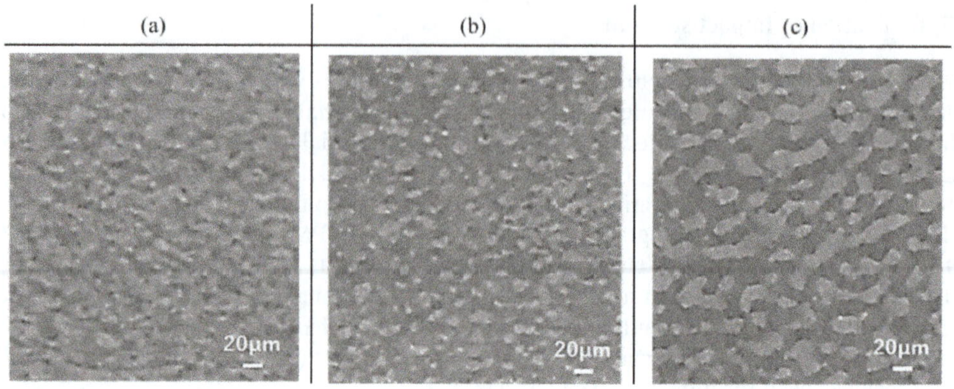

Figure 5: Morphology of materials from each extrusion method. (a) Distributive extrusion; (b) Elongational extrusion; and (c) Shear extrusion.

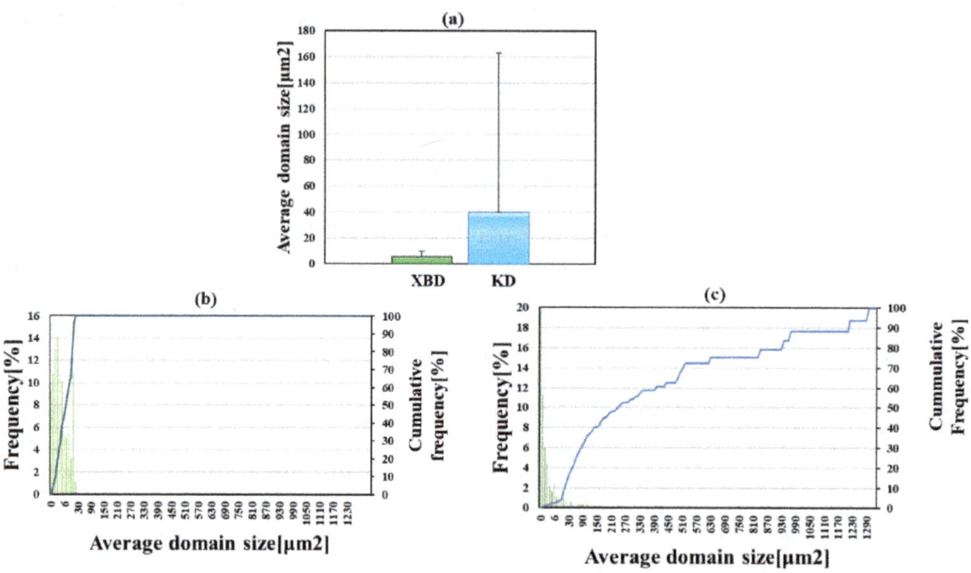

Figure 6: Differences in dispersed phase due to different extrusion methods. (a) Average domain size (average and standard deviation of measurements); (b) XBD; and (c) KD frequency distribution and cumulative frequency distribution.

$$N(r) = k \cdot r^{-D}. \tag{3}$$

In eqn (3), D is the fractal dimension and k is a constant. The box-counting method is a method of measuring $N(r)$ for each r by changing the value of r, plotting the relationship between $r$and $N(r)$ on a log–log graph, and obtaining D from its slope [12]. The observed image was 390 × 425 pixels, and the box was analysed by setting r to 4, 8, 16, 32, and 64. As a result, D of the KD sample was 1.698, and that of the XBD sample was 1.622, showing no significant difference in distributivity.

4.2 Impact strength

4.2.1 Comparison of impact strength

The impact strength results obtained from the Izod impact test are shown in Fig. 7. These results indicate that no significant improvement in impact strength was observed for any composition ratio. The impact strength mechanism of the PMMA/Bio-PC system is thought to involve the formation of shear bands when stress is applied [2]. Specifically, stress is concentrated at the interfaces of the dispersed particle, causing shear yielding of the matrix resin, resulting in the formation of shear bands that absorb impact energy, thereby enhancing impact strength [3]. Therefore, finer dispersion of the dispersed phase increases the number of dispersed particles, leading to a higher occurrence of shear bands, which improves energy absorption and ultimately enhances impact strength. When a load is applied to a PMMA/Bio-PC alloy, a stress concentration occurs at the tip of the plastic deformation region of the matrix surrounding the dispersed particles. When this stress concentration exceeds the crazing stress σ_c of the matrix resin, crazing form between the particles. As the load increases, these crazes grow into cracks, eventually leading to abrupt failure and acting as a barrier to increased impact strength. There is a critical interparticle distance d_l between dispersed particles at which cracking occurs, and cracking occurs between particles when the distance between dispersed particles exceeds this distance. d_l is shown in eqn (4) [5]:

$$d_l = 2R\exp\left\{\frac{\frac{\sigma_c}{\tau_0}-1}{2}\right\}. \tag{4}$$

In eqn (4), d_l is the critical interparticle distance, R is the average particle radius, σ_c is the crazing stress of the resin, and τ_0 is the shear yield stress of the resin. To prevent crack initiation, it is necessary to keep all interparticle distances below d_l through uniform distribution of the dispersed phase. In the next subsection, a uniform distribution model is developed, and morphological observations are made to study the distribution and dispersion aspects.

4.2.2 Uniform distribution model

A uniform distribution model was created for each composition, and the model's interparticle distance was compared with the critical interparticle distance. As shown in Fig. 8, the uniform

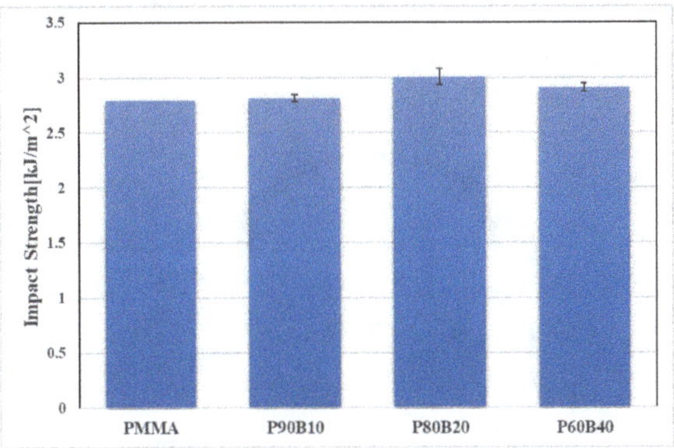

Figure 7: Impact strength for PMMA/Bio-PC weight composition ratio.

distribution model considers the case where a cubic lattice with a side d containing a particle at the center is laid out in a green cube with a side of 10 μm. In this case, the distance between the centers of the particles can be expressed by d. When the weight ratio of PMMA/Bio-PC is 90/10 wt%, the volume ratio is 91.1%/8.9%, so the volume V_B of Bio-PC in the cube V $=1000\mu m^3$ is:

$$V_B = 1000 \times 0.089 = 89\ [\mu m^3]. \tag{5}$$

If the radius of the dispersed particles is R, the volume V_P per dispersed particle can be expressed as follows:

$$V_P = \frac{4\pi R^3}{3}\,[\mu m^3]. \tag{6}$$

The number of dispersed particles that can exist in the cube can be expressed as eqn (7), considering eqns (5) and (6):

$$\frac{V_B}{V_P} = 89 \times \frac{3}{4\pi R^3} = \frac{267}{4\pi R^3}. \tag{7}$$

In the case of uniform distribution, the volume V_N of the cubic lattice that a single dispersed particle enters is:

$$V_N = V \times \frac{V_P}{V_B} = V \times \frac{4\pi R^3}{267} = \frac{4000}{267}\pi R^3\,[\mu m^3]. \tag{8}$$

Since the length of a side of the cubic lattice containing a dispersed particle can be considered to be equal to the distance between the dispersed particles, the distance d_{P1} between dispersed particles can be obtained from eqn (8) as follows:

$$d_{P1} = \sqrt[3]{V_N} = \sqrt[3]{\frac{4000}{267}\pi}\ R \fallingdotseq 3.61R. \tag{9}$$

Considering PMMA/Bio-PC weight composition ratio of 80/20 wt% and 60/40 wt%, $d_{P2} = 2.85R$ and $d_{P3} = 2.25R$, respectively.

Substituting the craze generation stress $\sigma_c = 68$MPa and the shear yield stress $\tau_0 = 39MPa$ [5] for the matrix resin into eqn (4), the critical interparticle distance d_l can be expressed as:

$$d_l = 2.90R. \tag{10}$$

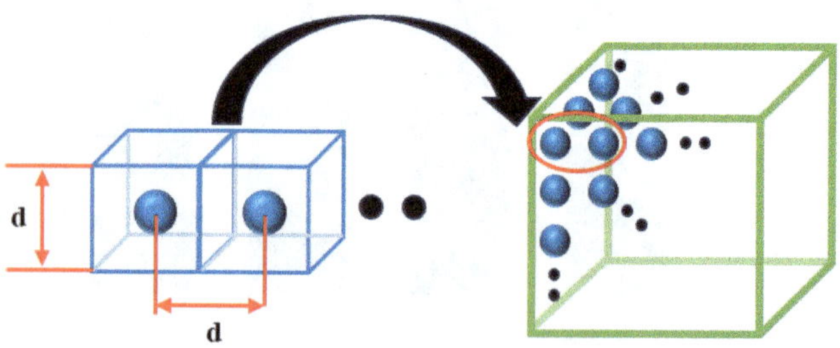

Figure 8: Uniform distribution model.

The results show that $d_{P1} > d_l$ for PMMA/Bio-PC = 90 wt%/10 wt%. Therefore, even with uniform distribution, the interparticle distance is always greater than the critical interparticle distance, and cracks are expected to occur and impact strength cannot be improved. On the other hand, when PMMA/Bio-PC = 80/20 wt% or 60/40 wt%, $d_P < d_l$, the interparticle distance is less than the critical interparticle distance when uniformly distributed. This result suggests that an increase in impact strength can be expected at a Bio-PC content of approximately 20 wt% or more. However, the impact strength was not improved at either 80/20 wt% or 60/40 wt%. This result is discussed by focusing on the phase dispersion and distribution from actual morphology.

4.2.3 Evaluation of phase dispersion and distribution by morphological observation

Observations of the dispersed phase were made using a field emission scanning electron microscope (JEOL Ltd. FE-SEM, JSM7001FD). The observation results shown in Fig. 9 indicate that all materials have a sea-island structure. However, the size of the dispersed phase in the PMMA/Bio-PC = 60/40 wt% case is not uniform, indicating that the dispersibility is poor. To visualize the dispersibility, the dispersed particle size and the number of dispersed particles were examined by analysis using ImageJ software. A graph of the number of dispersed particles is shown on Fig. 10. Fig. 10(a) shows the number of dispersed particles, and Fig. 10(b) shows a frequency graph of the dispersed particle diameter. The maximum number of dispersed particles was 1207 for PMMA/Bio-PC = 80/20 wt%, and the minimum was 139 for PMMA/Bio-PC = 60/40 wt%. From the frequency graph of the particle size distribution, it can be seen that the 80/20 wt% sample has the highest number of particles with a diameter of 0.50 μm or less, and that the variation in particle size distribution is also the smallest for the 80/20 wt% sample. It can be seen that the 60/40 wt% sample has the lowest number of particles with a diameter of 0.50 μm or less, and that the variation is the largest. These results indicate with an 80/20 wt%, the particle size can be reduced while controlling the variation in particle size, and the dispersion is good. On the other hand, with a 60/40 wt%, it is believed that it is difficult to reduce the dispersion phase and increase the number of particles. Furthermore, from the observation images, it can be seen that the large aggregates do not have a circular shape, but an irregular shape. From these results, it can be concluded that the dispersion is poor. This is thought to be due to the high content of Bio-PC. When the Bio-PC content is 40 wt%, even when an elongation kneading process is used, it is thought that dispersion is difficult, although it has on an island-in-the-sea structure. This suggests that a high content of Bio-PC may not be suitable for improving the impact resistance properties.

(a) (b) (c)

Figure 9: Morphological observation images for each composition ratio. (a) PMMA/Bio-PC = 90/10 wt%; (b) 80/20 wt%; and (c) 60/40 wt%.

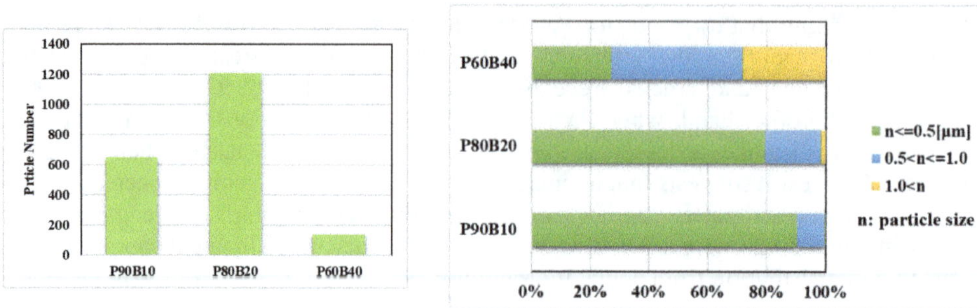

(a) (b)

Figure 10: (a) Number of dispersed particles for each composition ratio; and (b) Particle size distribution for each composition ratio.

The distance between particles was measured for PMMA/Bio-PC 80/20 wt% and 60/40 wt%. It was assumed that the distance between particles was the distance between the centres of gravity. This was obtained by analysing the SEM images in Fig. 9 and estimating the centre of gravity coordinates using ImageJ software. For the 80/20, the average particle diameter was extracted, and for the 60/40, because there was a lot of variation, the particle closest to the most frequent value was selected, and then three particles close to the target particle were selected and the distance was measured, after which the average of the three was taken.

The distance between the dispersed particles and the critical particle distance was then compared. Substitution of 0.221 μm, which is the average dispersion phase radius for 80/20 wt%, into eqn (4) yielded a critical particle distance of 0.640 μm. Similarly, for 60/40 wt%, the critical particle distance was determined to be 1.04 μm. The relationship between the measured particle distance and the critical particle distance is illustrated in Fig. 11. The bar graph contains the critical particle distance and symbols for the measured particle distance. The results show that for the 80/20 wt% mixture, approximately 60% of the measured particle distances exceeded the critical particle distance. From these results, it is

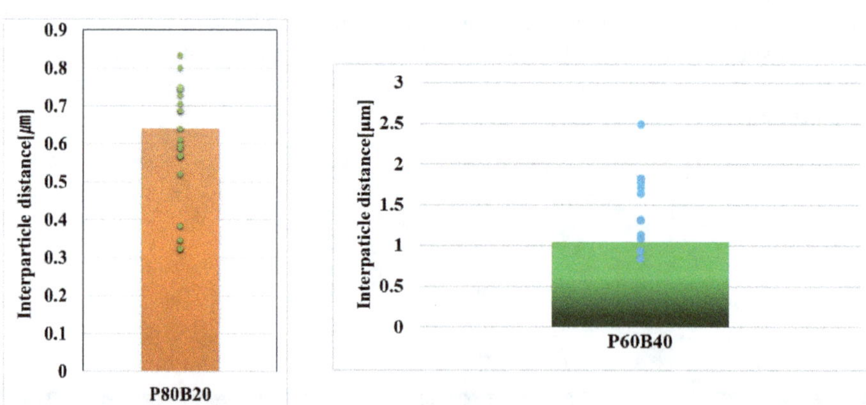

(a) (b)

Figure 11: Relationship between critical interparticle distance and measured interparticle distance for PMMA/Bio-PC. (a) 80/20 wt%; and (b) 60/40 wt%.

thought that the occurrence of cracks due to poor distribution when impact is applied inhibits the development of impact resistance. In the 60/40 mixture, approximately 83% of the particles exceed the critical particle-to-particle distance. These results suggest that increasing the content of Bio-PC to 40 wt% or more will worsen dispersion and distribution, making it difficult to impart impact resistance. Based on the above, it is thought that the optimal content for designing impact resistance in PMMA/Bio-PC composite materials using elongational flow is between 20 wt% and 40 wt%. However, this study has shown that distribution is still poor.

Based on the results of this study, it is necessary to look for optimal conditions that also improve the distribution to achieve higher impact strength. One factor that the effects distribution properties is the residence time in the twin-screw extruder [13]. By changing the operating conditions of the twin-screw extruder, it is possible to control the residence time of the molten resin. While it is believed that it is possible to improve the distribution by these operations, it is believed that it is necessary to consider the effect of changes in operating conditions on the elongation effect. Changes in flow near the XBD may affect the elongation effect and change the distribution. Although it is difficult to visualize the flow condition near the XBD, there is a correlation between the elongation stress and the pressure loss before and after the XBD, and it is supposed that it is possible to understand the elongation effect by obtaining the pressure from the experiment. The elongation stress and pressure loss can be related by the following formula [14]:

$$\sigma_E = {}^{3}\!/_{8}\,(n+1)\Delta P \quad 3-2-3-1, \tag{11}$$

where σ_E is the average elongation stress, n is the non-Newtonian index, and ΔP is the pressure loss. From this equation, it is possible to indirectly evaluate the effect of changes in operating conditions on the elongation effect by measuring the pressure. In the future, it will be necessary to establish the optimal values for the elongation effect and residence time required for dispersion and distribution to enhance the impact strength, with the amount of Bio-PC added between 20 wt% and 40 wt%.

5 CONCLUSION

In this study, the effect of elongation on PMMA/Bio-PC composites was investigated, with a comparison being made between the dispersibility and distributivity of composites produced by conventional shear mixing using KD and elongation mixing using XBD. In the case of elongation mixing using XBD, there was no significant difference in distributivity was observed when compared to shear mixing, however the refinement of the dispersed phase in terms of dispersibility was confirmed. Subsequently, the effect of the composition ratio of PMMA/Bio-PC on impact strength was measured, using XBD-based elongation mixing. The study revealed that there was no enhancement in impact strength across all composition ratios. However, although the need to improve the dispersibility remains, it was estimated that the optimal content range of Bio-PC for designing PMMA/Bio-PC composite materials with improved impact resistance through elongation flow is between 20 wt% and 40 wt%.One factor affecting the phase distribution is the residence time in the twin-screw extruder. By changing the operating conditions of the twin-screw extruder, it is possible to control the residence time of the molten resin. Whilst it is believed that improvements can be made to the distribution through these operations, it must be considered the effect of changes in operating conditions on the elongation effect. In future research, it will be necessary to establish the optimal values for the elongation effect and residence time required for dispersion and distribution to enhance the impact strength.

ACKNOWLEDGEMENTS

I would like to express my gratitude to Professor Tanaka and Professor Sasada for providing me with various opportunities and valuable advice throughout this research. I would also like to extend my appreciation to Ms Birei Tanaka and Ms Yuki Masaki for their invaluable cooperation as collaborative researchers.

REFERENCES

[1] Yoshida, M., Development and prospects of bioplastics. *Japan Society of Material Cycles and Waste Management (JSMCWM)*, **30**(2), pp. 123–130, 2019.

[2] Narisawa, I., Ishikawa, Y., Ishikawa, T. & Ogawa, H., Fracture and crazing of polymers under plane strain. *Polymer Journal*, **36**(8), pp. 543–556, 1979.

[3] Narisawa, I. & Kuriyama, T., Impact and fatigue properties of polymer blends. *Journal of the Society of Rubber Industry, Japan*, **68**(12), pp. 875–882, 1995.

[4] Hayashi, T., Watanabe, T., Tanaka, H. & Toshio, N., Morphology and mechanical properties of three-component immiscible polymer alloys. *Polymer Journal*, **49**(4), pp. 373–382, 1992.

[5] Ishikawa, Y., Morphology design for impact strength improvement. *Polymer Journal*, **40**(10), pp. 680–683, 1991.

[6] Hisamatsu, T., Nakamo, A., Ishikawa, Y. & Iwakura, K., Analysis of stress concentration relaxation mechanisms in polymer alloys by plastic mechanics. *Seikei-Kakou (Journal of the Japan Society of Polymer Processing)*, **11**(3), pp. 187–193, 1989.

[7] Tanaka, T. & Matsumoto, K., Kneading and dispersion technology of nanocomposites and polymer blends utilizing extensional flow in twin-screw extruders. *Advanced Functions of Plastics and Multiphase Flow*, **38**(2), pp. 114–121, 2024.

[8] Kajiwara, T., Theory and computer simulation of melt mixing for polymeric materials. *Proceedings of the 60th National Congress of Theoretical and Applied Mechanics*, 2011.

[9] Matsumoto, K. & Tanaka, T., A modified blister mixing element for generating extensional flow in a twin-screw extruder: Process characterization and dispersion state of carbon nanotubes in cyclo-olefin polymer. *Polymer Engineering and Science*, **62**(4), pp. 1223–1238, 2022.

[10] Grace, H.P., Dispersion phenomena in high viscosity immiscible fluid systems and application of static mixers as dispersion devices in such systems. *Chem. Eng. Commun.*, **14**(3–6), pp. 225–277, 1981.

[11] Tokihisa, M., Yakemoto, K., Sakai, T., Utracki, L.A., Sepehr, M., Li, J. & Simard, Y., Extensional flow mixer for polymer nanocomposites. *Polym. Eng. Sci.*, **46**, pp. 1040–1050, 2015.

[12] Fujimoto, K., Yamamoto, T. & Shioya, T., Evaluation of fractal dimensions of fracture surfaces using box-counting method. *Mechanical Engineering Congress*, **6**, pp. 66–67, 2006.

[13] Hirofumi, T., Mikata, Y., Matsuda, S. & Atsushi, M., Dispersion of carbon-nanotubes in a polymer matrix by a twin-Screw extruder. *Journal of the Society of Plastic Engineering Japan*, **14**(2), pp. 126–131, 2002.

[14] Matsumoto, K., Study on dispersive mixing for nanocomposite materials with co-rotating twin-screw extruder. PhD Dissertation, Doshisha University, 2017.

SECTION 2
MATERIAL CHARACTERISATION

STRENGTH PROPERTY EVALUATION OF SURFACE-TREATED NYLON FABRICS FOR WATERPROOF PRODUCTS AFTER SEAM PROCESSING

HIDEAKI KATOGI
Department of Human Environmental Sciences, Jissen Women's University, Japan

ABSTRACT
Many companies have announced waterproof clothing, cover sheets, fishing wear and other products. Functionality of the constituent materials of waterproof products is extremely important for comfort. However, the manufacture of products using waterproof fabrics must be conducted using seam treatments. Therefore, this study examined seam method effects on strength properties of waterproof treated nylon fabrics for sustainable development goals. Fabric surfaces were treated using polyvinyl chloride (PVC) processing. Evaluation of waterproof fabric contact angles was conducted before seam tensile testing. The seam treatment was straight with embroidered seams. Seam tensile tests of PVC-treated nylon fabrics were conducted after those of straight and embroidered seams. Polyester yarn sewing thread was used. The environment temperature was room temperature. The crosshead speed was 100 mm/min. The following results were obtained. The contact angle on the surface of PVC-treated nylon fabric was about 86°. Load–displacement curves suggested that the fracture toughness of PVC-treated nylon fabric with an embroidered seam was much larger than that of PVC-treated nylon fabric with a straight seam. Their results implied that the fracture toughness of PVC-treated nylon fabric was affected mainly by the seam method.
Keywords: clothing, embroidered seam, fracture toughness, load–displacement curve, straight seam.

1 INTRODUCTION

Fabric sewing processes are applied to products using sewing machines. Stitch method effects on the mechanical properties of composite materials have been investigated for improvement of new products. In fact, many reports have described mechanical properties of non-stitched and stitched composite materials [1]–[11].

Ai et al. [12] reported some details about mechanical properties of stitched sandwich panel structures. The stitching parameters have strengths and failure features of the stitched sandwich panel structure. Tapullima et al. [13] reported characterization of a mode II specimen using I-fibre stitching processing. The cohesive law for mode II fracture mechanisms was affected by stitching parameters. Song et al. [14] reported some effects of stitching yarn types on the flexural fatigue properties of 3D-stitched carbon fibre composites. The flexural fatigue life (10^6 cycles) of 3D-stitched carbon fibre composites continued even when the maximum fatigue stress was 542 MPa.

Moreover, many reports have described mechanical properties of fibres, yarns, and fabrics for application to new fibre products [15]–[18].

Regarding new products, Katogi et al. [19] reported the biodegradation of coffee with added regenerated cellulose fibre designed to advance sustainable development goals. The coffee with added regenerated cellulose fibre exhibited good performance. Singal et al. [20] reported mechanics in knitted materials using stitching. Finite element analysis of the knitted material was conducted for prototype models of fibre products, revealing the deformation mechanics in the knitted fabric constituent material.

Clothing is used occasionally underwater and in moist environments. Waterproof fabrics are used outdoors for plumbing work and similar tasks. Functionalities of the waterproof products' constituent materials are extremely important for comfort. Generally, tensile

WIT Transactions on The Built Environment, Vol 216, © 2025 WIT Press
www.witpress.com, ISSN 1743-3509 (on-line)
doi:10.2495/HPSM250031

properties of constituent materials are decreased by water absorption. Water and moisture absorption affect some constituent materials of the product. Therefore, the clothing should be investigated for its mechanical properties when using a waterproof fabric as a constituent material. Moreover, the manufacture of products using waterproof fabrics must be done using seam treatment. Nevertheless, few papers have described seam method effects on mechanical properties of fabrics used for waterproofing of clothing.

For this study, the strength properties of surface treated nylon fabrics were evaluated for waterproof products after seam processing.

2 MATERIAL AND SEAM METHOD

The material is PVC-treated nylon fabric. Fig. 1 shows the PVC-treated nylon fabric surface. The constituent material of the yarn was polyester yarn for sewing processing. Fig. 2 portrays the surfaces of straight and embroidered-seam nylon fabrics. The sewing process was conducted using a sewing machine (HZL-J900R; Juki Corp.).

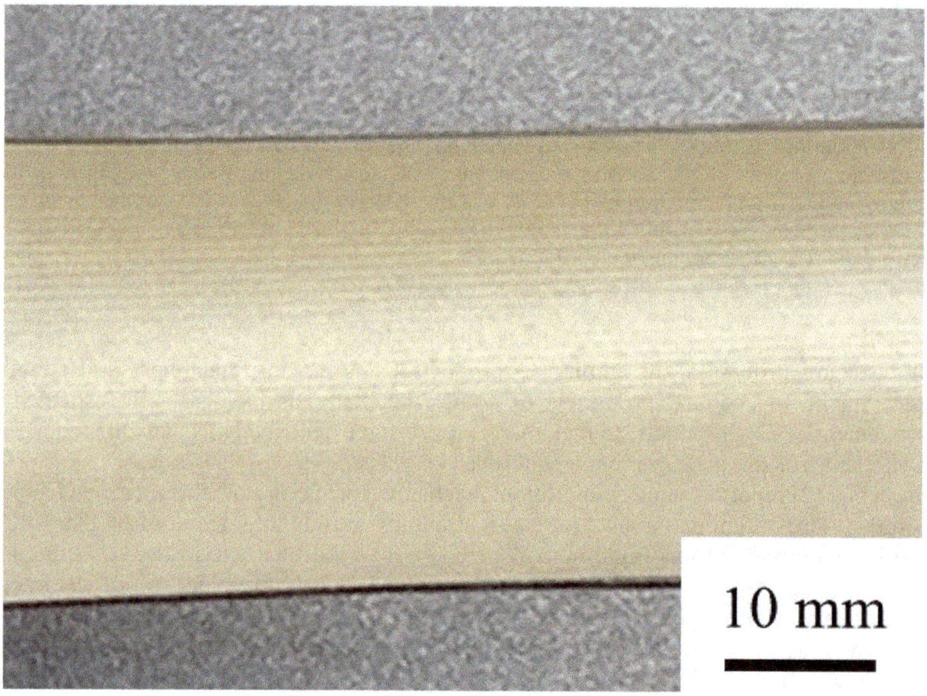

Figure 1: Surface of PVC-treated nylon fabric (front side).

3 TESTING METHOD

Seam tensile tests of PVC-treated nylon fabrics with straight and embroidered seams were conducted based on Japanese Industrial Standard (JIS) L 1093. The crosshead speed for static tensile testing was 100 mm/min. The gauge length was 100 mm. The environment temperature was room temperature. A testing machine for PVC-treated nylon fabrics with straight and embroidered seams was used (EMT-5kNV-50; Shimadzu Corp.). The seam allowance length was 10 mm. Fig. 3 presents a schematic drawing of the PVC-treated nylon fabric during seam tensile tests.

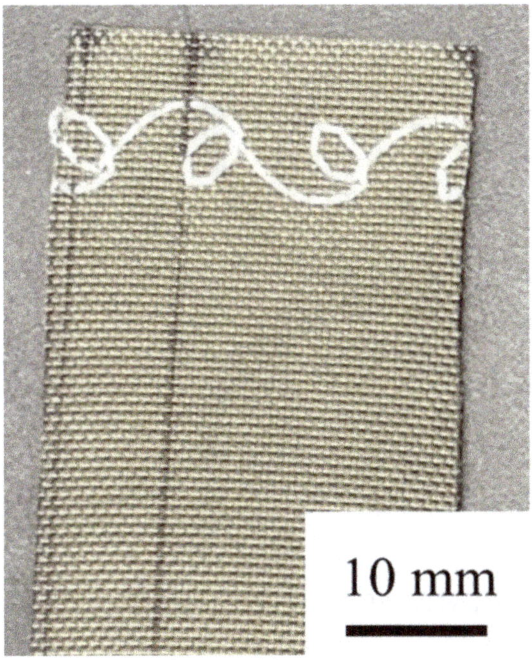

(a)

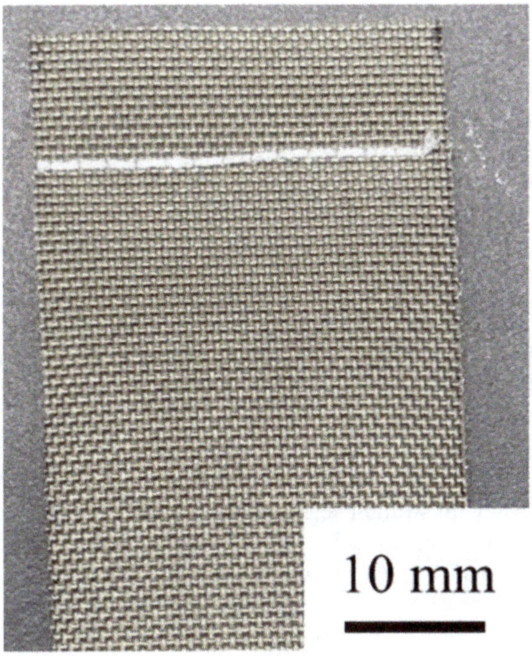

(b)

Figure 2: Surfaces of PVC-treated nylon fabrics. (a) Straight seams; and (b) Embroidered seams.

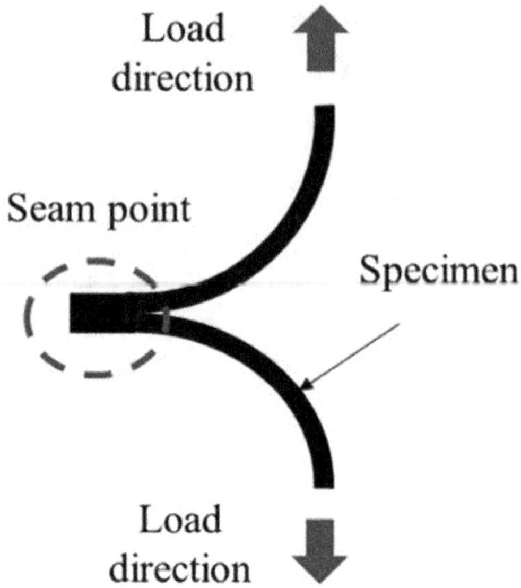

Figure 3: Schematic drawing of PVC-treated nylon fabric during seam tensile testing.

4 RESULTS AND DISCUSSION

4.1 Contact angle of nylon fabric

Fig. 4 presents contact angles on the surfaces of PVC-treated nylon fabrics for the front and back sides. The contact angles on surfaces of the nylon fabrics of the front and back sides were, respectively, approximately 86° and 70°. The nylon fabric surface at the front side had undergone using PVC treatment. The water absorption of PVC as a constituent material almost did not occur because water absorption of the PVC at 80°C was about 0.3% [21]. Therefore, the PVC treatment affected the nylon fabric waterproofing property.

4.2 Seam strength property

Fig. 5 shows load–displacement curves of PVC-treated nylon fabrics with straight and embroidered seams. Load–displacement curves of PVC-treated nylon fabrics with straight and embroidered seams were displaced approximately 3 mm during seam tensile testing. Subsequently, load–displacement curves of PVC-treated nylon fabrics with straight and embroidered seams became linear. After maximum loading at the breaking point, the load of PVC-treated nylon fabric with straight seams decreased slowly with increased displacement. However, the load of PVC-treated nylon fabric with an embroidered seam showed great variation with an increase of displacement after load at the breaking point. Maximum loads of PVC-treated nylon fabrics with straight and embroidered seams were, respectively, approximately 94 N and about 276 N. Displacements at breaking points of PVC-treated nylon fabrics with straight and embroidered seams were, respectively, about 13 mm and about 33 mm. The fracture toughness of PVC-treated nylon fabric with embroidered seams was greater than that of PVC-treated nylon fabric with a straight seam. These results imply that the seam strength properties of PVC-treated nylon fabrics were affected by the seam method using polyester yarn.

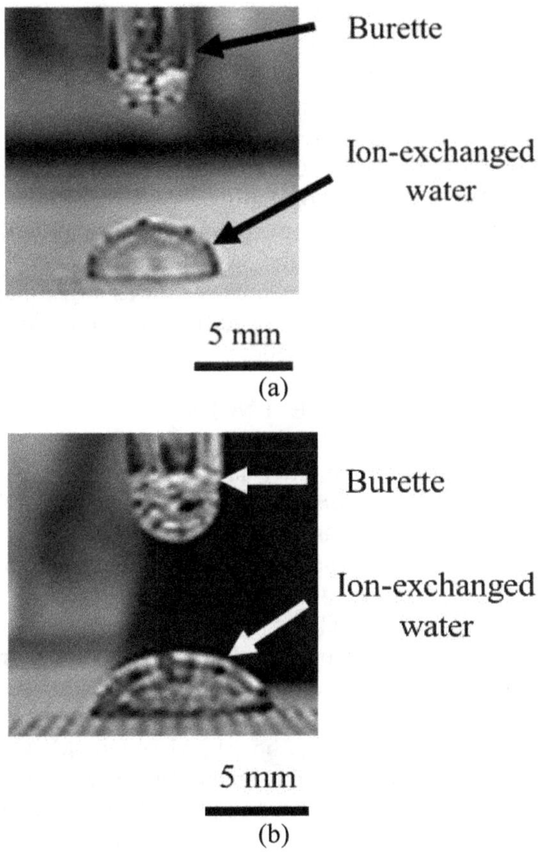

Figure 4: Measurements of contact angles of PVC-treated nylon fabrics. (a) Front side; and (b) Back side.

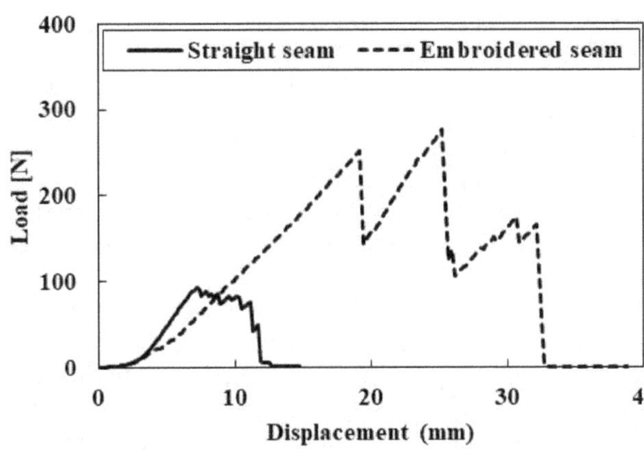

Figure 5: Load–displacement curves of PVC-treated nylon fabrics (straight and embroidered seams).

5 CONCLUSIONS

This study evaluated strength properties of surface-treated nylon fabrics designed for waterproof products after seam processing. The following conclusions were obtained.

The contact angle on the nylon fabric surface at the front side was greater than that of the nylon fabric at the back side because the nylon fabric surface at the front side had undergone PVC treatment. The maximum load of PVC-treated nylon fabric with an embroidered seam was higher than that of PVC-treated nylon fabric with a straight seam. Furthermore, the displacement at the breaking point of PVC-treated nylon fabric with the embroidered seam was greater than that of PVC-treated nylon fabric with a straight seam. The fracture toughness of PVC-treated nylon fabrics with the embroidered seam was larger than that of PVC-treated nylon fabrics with a straight seam. Therefore, seam strength properties of PVC-treated nylon fabric were affected mainly by the seam method.

REFERENCES

[1] Tan, K.T., Watanabe, N., Iwahori, Y. & Ishikawa, T., Effect of stitch density and stitch thread thickness on compression after impact strength and response of stitched composites. *Composites Science and Technology*, **72**, pp. 587–598, 2012.

[2] Yudhanto, A., Watanabe, N., Iwahori, Y. & Hoshi, H., Effect of stitch density on tensile properties and damage mechanisms of stitched carbon/epoxy composites. *Composites Part B: Engineering*, **46**, pp. 151–165, 2013.

[3] Matsumoto, K., Nagasaka, T., Takemura, K. & Tanaka, T., Influence of nanofiber loading and moulding conditions on the joining strength of thermoplastic composites fabricated by injection over-moulding process. *WIT Transactions on the Built Environment*, vol. 196, WIT Press: Southampton and Boston, pp. 113–124, 2020.

[4] Sun, C., Jia, P., Chen, C., Moradi, A., Zhou, J., Al Teneiji, M., Cantwell, W.J. & Guan, Z.W., The effect of carbon fibre stitching on the tensile behaviour of secondary bonded single-and double-lap composite joints. *Composite Structures*, **265**, 113774, 2021.

[5] Li, H., Yu, Y., Xu, X., Chen, T. & Lu, W., Enhancing the fracture toughness of laminated composites through carbon nanotube belt stitching. *Composites Science and Technology*, **204**, 108632, 2021.

[6] Katogi, H. & Tsunekawa, H., Short-term outdoor exposure effects on cotton fabric abrasion properties. *WIT Transactions on Engineering Sciences*, vol. 133, WIT Press: Southampton and Boston, pp. 47–54, 2021.

[7] Matsumoto, K., Itabashi, M., Kawasumi, A., Takemura, K. & Tanaka, T., Selective reinforcement of joining interface using nanofibers in single-lap joints of thermoplastic composites fabricated by the injection overmolding process: Creep deformation behaviour. *WIT Transactions on the Built Environment*, vol. 209, WIT Press: Southampton and Boston, pp. 3–14, 2022.

[8] Murayama, R., Obunai, K. & Okubo, K., Effect of sub-micron glass fiber addition on CAI strength of UD-CFRP. *International Journal of Computational Methods and Experimental Measurements*, **10**, pp. 271–281, 2022.

[9] Katogi, H., Tensile property of jute fiber sliver reinforced green composite material under loading speed. *Key Engineering Materials*, **958**, pp. 187–192, 2023.

[10] Katogi, H., Effect of loading speed on tensile property of single flax yarn/shellac green composite material. *Materials Science Forum*, **1134**, pp. 33–38, 2024.

[11] Matsumoto, K., Takemura, K., Kitamura, R., Katogi, H., Tanaka, T. & Takagi, H., Cellulose nanofiber-introduced continuous-ramie yarn-reinforced polylactic acid filament for 3D printing: Novel fabrication process and mechanical properties. *Composites Part A: Applied Science and Manufacturing*, **176**, 107836, 2024.

[12] Ai, S., Chen, Y. & He, R., Effects of stitching parameters on the mechanical properties of stitched sandwich panel structures. *Ceramics International*, **47**, pp. 5249–5255, 2021.

[13] Tapullima, J., Song, S.H., Kweon, J.H. & Choi, J.H., Characterization of mode II specimen using I-fiber stitching process. *Composite Structures*, **255**, 112863, 2021.

[14] Song, W., Liu, T., Song, C., Wang, S., Jing, Y., Ma, J., Li, G., Zhou, R., Dou, H. & Fan, W., Effects of stitching yarn types on flexural fatigue properties of 3D stitched carbon fiber composites. *Polymer Testing*, **115**, 107744, 2022.

[15] Jevšnik, S. & Žunič-Lojen, D., Drape behaviour of seamed fabrics. *Fibers and Polymers*, **8**, pp. 550–557, 2007.

[16] Abrishami, S., Ezazshahabi, N. & Mousazadegan, F., Assessment of the influence of stitching on the tensile stress relaxation of laminated fabrics. *Journal of Industrial Textiles*, **51**, pp. 969S–989S, 2022.

[17] Li, H., Zhang, R., Min, S., Zhou, Y. & Sun, J., Parametric study on the ballistic performance of seamed woven fabrics. *Defence Technology*, **24**, pp. 173–189, 2023.

[18] Wu, M., Wei, L., Chen, H., He, G., Tang, N., Wu, M. & Liu, S., Characterization of tensile properties of chain stitch structure fabric with negative Poisson's ratio yarn. *The Journal of the Textile Institute*, pp. 1–6, 2024.

[19] Katogi, H., Tsunekawa, H. & Takata, A., Biodegradation of cuprammonium rayon with added coffee grounds produced by wet spinning. *Materials Today: Proceedings*, **42**, pp. 161–164, 2021.

[20] Singal, K., Dimitriyev, M.S., Gonzalez, S.E., Cachine, A.P., Quinn, S. & Matsumoto, E.A., Programming mechanics in knitted materials, stitch by stitch. *Nature Communications*, **15**, p. 2622, 2024.

[21] Hasegawa, S., Effect of water absorption on surface properties of polymer insulating material. *Research Reports of Akita National College of Technology*, **35**, pp. 42–46, 2000.

HIGH AND LOW CYCLE FATIGUE OF A 6060 ALUMINIUM ALLOY

LORENZO PAGLIARI[1], RICCARDO GEROSA[2], DAVIDE PANZERI[2] & FRANCO CONCLI[1]
[1]Faculty of Engineering, Free University of Bolzano, Italy
[2]Dipartimento di Meccanica, Politecnico di Milano, Italy

ABSTRACT

Lightweight alloys are gaining significant interest in light of their high strength-to-weight ratio, which paves the way to more efficient systems with a direct impact on pollutant emissions, sustainability and carbon footprint. The results of a High- (HCF) and Low-Cycle-Fatigue (LCF) characterization testing campaign performed on a 6060-Aluminium alloy are presented. The 6XXX-series aluminium alloys features Magnesium and Silicon. These alloying elements form Magnesium Silicide (Mg2Si) as strengthening phase. The two predominant components must maintain perfect proportions to create the Al-Mg-Si system. 6XXX-series alloys are most frequently used for extrusion, as they provide a good balance of strength, corrosion resistance, weldability, machinability, and surface finish. Understanding the exact mechanical behaviour is crucial for its large-scale adoption. A series of cylindrical samples, manufactured with traditional techniques, was used to determine the High-Cycle-Fatigue properties of the 6060-alloy exploiting the statistical approach proposed by Dixon. The maximum number of cycles (run-out) was set to 5000000. The finite-life range was also studied by means of additional tests at higher stress levels. Regarding the LCF, tests were carried out with a strain-ratio equal to -1. The fatigue data were studied by means of the Coffin-Manson-equation and the cyclic and static behaviours were compared. The 6XXX-Aluminum series do not normally exhibit elastic unstable fracture. These alloys are so tough that fracture toughness is rarely a design criterion. However, the present set of HCF samples exhibited intergranular fracture. This phenomenon occurs when cracks propagate along the grain boundaries when these are weakened, typically due to adverse environmental conditions or sensitization phenomena. In light of this evidence, the results were compared with data from literature. While the mechanical properties remain within the typical ranges, they are very low. Further analyses have been conducted to explain such poor mechanical behaviour, through the very unusual fracture mode that was observed.

Keywords: aluminium, high-cycle-fatigue, low-cycle-fatigue, intergranular fracture.

1 INTRODUCTION

The growing demand for increasingly high-performance systems has paved the way for the large-scale use of lightweight materials. Among these, aluminium is probably the most used, showing a constant growth throughout the years. Already in year 2000, aluminium alloys had overtaken plastic in the most-used-material ranking, becoming the third most widespread material in automobiles manufacturing. The success of aluminium alloys is mostly related to its extremely efficient production process.

The widespread and effective adoption of a material for new optimized components is strongly related to a deep knowledge of their properties. Apart from applications where gross yielding or high ductility fracture represents a limiting failure condition, tensile strength and yielding are insufficient requirements to design fracture-resistant structures. In addition to this, a separate chapter could be spent when dynamic loading is considered.

Achieving the project requirements necessitates considering severe stress concentrators, for which possible failure mechanisms may include high-cycle and low-cycle fatigue, brittle fracture, or even fracture from some combination of cyclic and static loading.

WIT Transactions on The Built Environment, Vol 216, © 2025 WIT Press
www.witpress.com, ISSN 1743-3509 (on-line)
doi:10.2495/HPSM250041

The proper alloy selection represents an important means of minimizing premature fracture in engineering structures but inevitably require a very deep knowledge of the material properties.

2 THE ALUMINIUM ALLOYS

Similar to other face-centred cubic materials, aluminium alloys usually do not experience a sudden ductile-to-brittle transition in fracture behaviour as temperatures decrease [1], [2]. Tensile test results show that nearly all aluminium alloys remain insensitive to strain rates ranging from 10^{-5} mm/mm/s to 1 mm/mm/s ($\approx 10^{-5}$ MPa/s) at both room and low temperatures [2]. Consequently, aluminium is well-suited for structural applications across a wide range of temperatures and loading conditions.

One possible classification of aluminium alloys is based on their strengthening mechanisms. Some alloys gain strength primarily through strain hardening (H), while others rely on solution heat treatment and precipitation aging (T). Another widely used classification, established by The Aluminium Association, identifies the primary alloying element by the first digit of the alloy designation.

The 1XXX series, for instance, is mostly pure Al, strengthened via cold working. The 2XXX series consists of a combination of Al-Cu-Mg, strengthened via heat treatments. The 3XXX series, is made up of Al-Mn-Mg (cold working), the 4XXX series consists of Al-Mg or Al-Mg-Mn (cold working), the 6XXX series consists of heat-treated Al-Mg-Si, the 7XXX series of heat-treated Al-Zn-Mg-(Cu), and the 8XXX series of heat treated Al-Li [3].

Among these, the 5XXX and 6XXX alloys provide medium-to-relatively high strength, excellent corrosion resistance, and exceptional toughness, making fracture toughness considerations generally unnecessary. The 6XXX series, focus of this work, is known for its good formability and weldability at moderate strengths and is widely applied in conventional structural projects. The typical range of tensile strength lies between 150 and 380 MPa.

3 6060-T6

This study focuses on a 6060 alloy (ISO designation AlMgSi). The EN-573-3:1994 foresee specific ranges for the alloying elements for the 6060 alloy. In particular, Si should be between 0.3 and 0.6, Fe between 0.1 and 0.3, Cu, Mn and Ti between 0 and 0.1, Mg between 0.35 and 0.6, Cr between 0 and 0.05 and Zn between 0 and 0.15.

For this specific alloy, limited data are available in literature. It usually undergoes a T6 heat treatment that foresees a heating process at very high temperatures (around 535°C), followed by a cooling step. The quenching is made in a liquid solution bath (usually water or glycol) at a temperature between 65°C and 100°C. After quenching, the castings are aged in a furnace, heated to 175°C for several hours and then allowed to cool naturally.

It has been shown that the T6 state increases the fatigue strength by approximately 30% compared to the T4 state [4]. Pisapia et al. [5] tested the 6060 aluminium alloy under different heat treatments under LCF loading and found that the 6060-T6 alloys exhibited high cyclic hardening behaviour. Hopperstad et al. [6] studied the uniaxial cyclic behaviour of the aluminium alloy 6060 in tempers T4 and T6. In the latter condition, the material exhibits cyclic softening. Comparison of corresponding hysteresis loops, shows that the effect of prior history is not very significant for temper T6. The scholars reported indications of the Bauschinger effect and the cyclic hardening in the form of the Chaboche model. Jha et al. [7] reported the properties of the 6060 aluminium alloy. The material was found to have high tensile strength (UTS = 374 MPa), yield strength (σ_Y = 303 MPa) and hardness. However, it is characterized by relatively low ductility.

In contrast, Fernandes et al. [8] reported quite low mechanical properties for the 6060-T6 alloy (σ_Y = 150 MPa, UTS = 190 MPa).

Despite these outliers, most of the records report values around 210 ÷ 230 MPa for what concerns the yielding and around 230 ÷ 250 MPa for that concern UTS.

Hockauf et al. [9] tested a 6060 alloy with different heat treatments. For its standard T6 version, the scholars reported a tensile strength of 233 MPa, and a yield strength of 208 MPa. Hockauf et al. reported also the high cycle fatigue properties in the form of Basquin coefficients (b = −0.0787, σ_f' = 354 MPa). The scholars stated also that the crack growth threshold ΔK_{th} lies at 1.5 MPa·m$^{1/2}$. Moreover, Hockauf et al. [10] determined the influence of initial deformation on fatigue life and the effect of aging before cyclic fatigue. The proper fatigue tests were performed with the control of plastic strain amplitudes in the range $\varepsilon_{apl} = 1\cdot10^{-3} \div 5\cdot10^{-3}$. In addition to the determined fatigue characteristics in the stress–strain system, the resulting fatigue fractures were subjected to fractographic analysis.

Borrego et al. [11] performed LCF tests on 6060-T6 alloys subjected to strain ranges between 0.32% and 4%. They report a tensile strength of 240 MPa, and a yield strength of 215 MPa. The scholars include in their publication also the cyclic properties in the form of the Ramberg–Osgood (RO) coefficients (k' = 267.3, n' = 0.38). The tested alloy shows a brief initial cyclic hardening stage followed by long-term cyclic softening. Beside this, also the fatigue characteristics were presented in the form of Basquin–Coffin–Manson (BCM) coefficients (b = −0.084, σ_f' = 376.5 MPa, c = −0.537, ε_f' = 0.157).

The static characteristics were measured also by Corre et al. [12] (σ_Y = 229 ÷ 235 MPa, UTS = 251 ÷ 259 MPa). The scholars observed both transgranular ductile fracture with dimples, as expected for an aluminium alloy at room temperature, and intergranular ductile fracture, which is less common. Other scholars have already reported this type of fracture surfaces for aluminium alloys with a mixed mode failure [13], especially in Al-Mg-Si alloys [14]. In their study, the scholars observed a homogeneous fracture surface with no distinct features, consistent with the smaller amount of main precipitates (Mg_2Si hardening particles) and the complete absence of dispersoid phase. The observed nearly ideal Masing behaviour is coherent with the low level of precipitates and compatible with the Christ and Mughrabi hypothesis [15] (i.e. an important requirement for Masing-type behaviour in multiple-phase materials is that the particle/dislocation interaction is of minor importance for plastic deformation compared to dislocation interactions).

Kowalski et al. [16], [17] carried out a fractographic analysis of fatigue fracture surfaces of aluminium alloys after cyclic bending and torsion tests. Crack propagation occurs mainly by connecting successive cracks that are formed at adjacent grain boundaries. It was found that the microstructure, type, and size of cyclic loads determine the share of trans-crystalline, quasi-split, and ductile cracks, thus determining the fractography of fatigue samples.

4 EXPERIMENTAL SETUP

High-cycle-fatigue (HCF) tests were conducted on cylindrical specimens (Fig. 1 (left)) designed following the ASTM E466 standard [18]. The experiments were carried out using a STEPLab-UD04 fatigue testing machine, which can apply loads of up to 5 kN at a maximum frequency of 35 Hz. It was used for previous studies on aluminium alloys [19]–[21] as well as other materials [22], [23].

Alongside the HCF tests, quasi-static tensile characterization was performed. The geometry of the specimens used in these tests is depicted in Fig. 1 (left). Tensile testing was conducted at room temperature (20°C) with a crosshead speed of 0.1 mm/min. The fractured specimens exhibited a cone-cup shaped fracture surface, which is characteristic of ductile

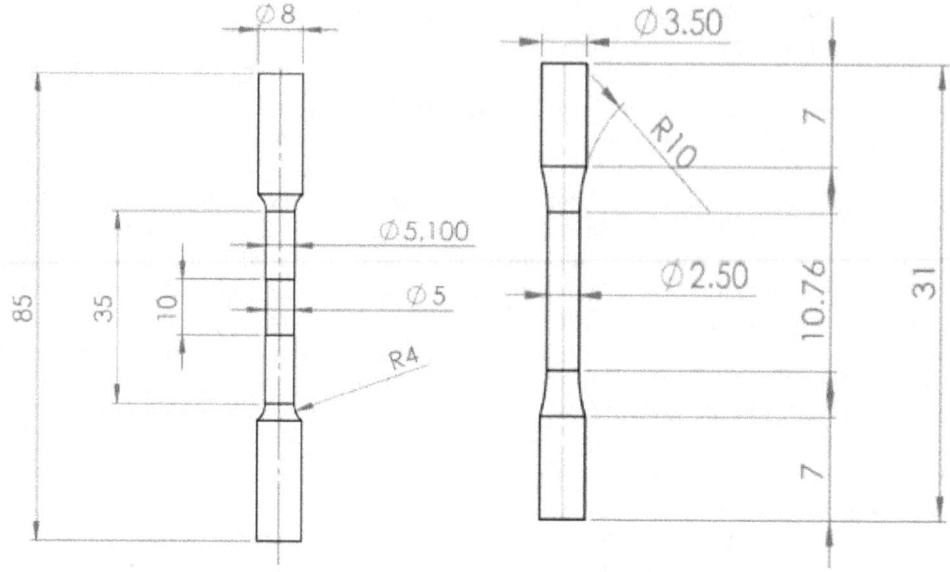

Figure 1: Standard specimen for quasi-static and HFC (left), according to ASTM E466 and LCF (right) tests according to ASTM E606.

failure. Additionally, plastic deformation was confirmed by the presence of a fibrous and irregular necking region.

Preliminary static tests were also carried out to determine the yield point (Fig. 2). The data obtained were subsequently used to calibrate the Basquin parameters, as detailed in the following sections.

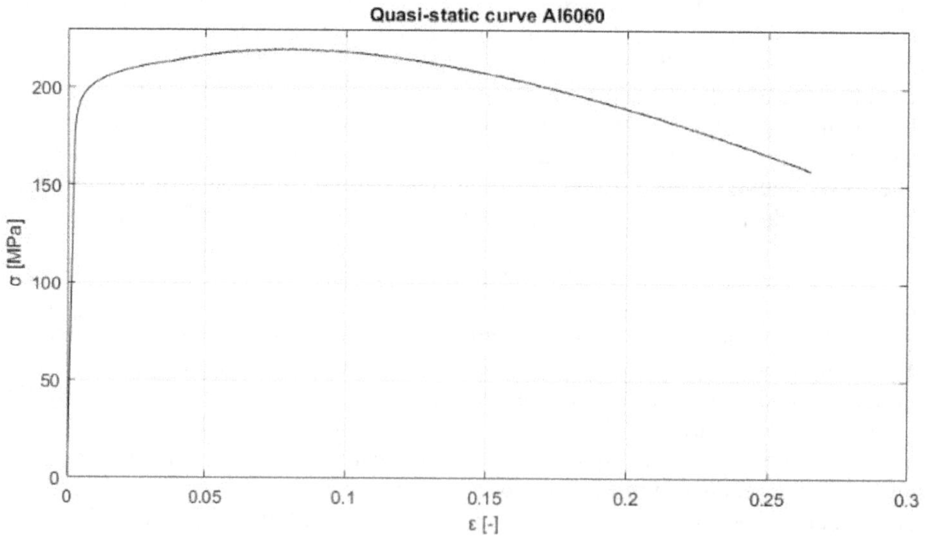

Figure 2: Quasi-static curve for the Al 6060.

4.1 HCF

HCF tests followed the ASTM E466 standard [18], and the fatigue limit was estimated using the short stair-case method [24]–[28]. Specifically, multiple tests were performed at varying load levels. This approach involves defining a force increment, which directly influences results accuracy. In this study, the increment was set at 50 N, corresponding to a stress interval ($\Delta\sigma$) of approximately 10.2 MPa.

If the initial test, conducted at a given force level, reached the run-out condition (5 million cycles without failure), the subsequent test was performed at a higher force level, increased by the defined increment. Conversely, if the specimen failed, the subsequent test was conducted at a lower force level, reduced by the same increment.

Fig. 3 illustrates the short stair-case sequences (five tests – the suggested value recommended by Dixon [24]) for the Al 6060 series. Unlike the conventional stair-case method, Dixon's statistical approach accounts not only for the number of failures (X) and run-outs (O) but also for the sequence order. A statistical correction coefficient ensures the reliability of the results.

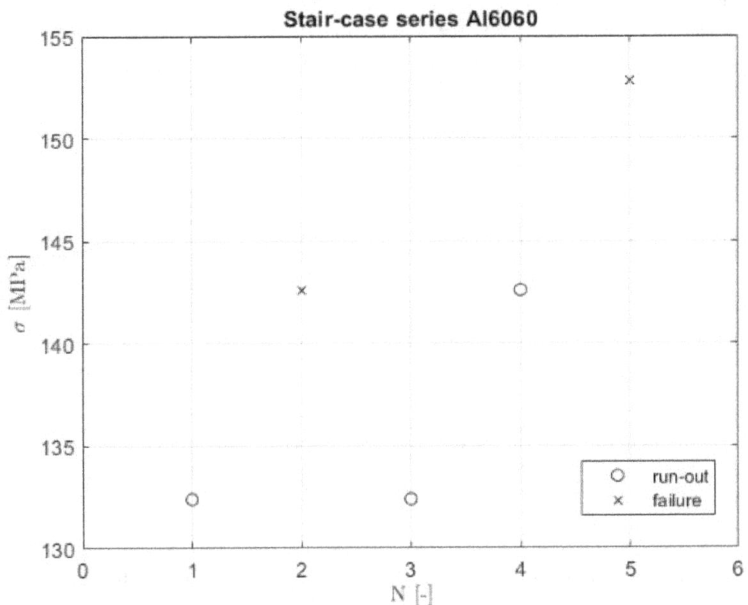

Figure 3: Results of the stair-case testing.

The fatigue limit was determined using the following equation:

$$\sigma_F = x_f + k \cdot \Delta\sigma, \tag{1}$$

where σ_F represents the fatigue limit at 50% probability, x_f denotes the stress level of the final test (ID = 5 in Fig. 3), and k is a sequence-dependent coefficient. The test sequence followed the order O-X-O-X-X, with a corresponding factor of −0.878. Based on this, the calculated fatigue limit was 143.8 MPa.

4.2 LCF

Low-cycle fatigue (LCF) tests were conducted on cylindrical specimens with standard dimensions in accordance with ASTM E606 [29], as depicted in Fig. 1 (right). Strain-controlled fatigue experiments were carried out using the STEPLab UD04 testing machine following the guidelines outlined in ASTM E606 [29]. The tests were performed at a specified strain ratio, with the testing frequency set at 0.1 Hz which represents the lowest frequency recommended by ASTM E606. This frequency remained constant throughout the tests, as increasing it could influence the results by slightly raising the sample temperature and altering its mechanical properties. The stabilized cycles for each deformation level were interpolated using the RO equation (eqn (2)) to derive the stabilized curves (Fig. 4).

$$\varepsilon_a = \varepsilon_{ae} + \varepsilon_{ap} = \frac{\sigma_a}{E} + \left(\frac{\sigma_a}{K'}\right)^{\frac{1}{n'}}. \tag{2}$$

In this equation, ε_a denotes the total strain amplitude, while ε_{ae} and ε_{ap} represent the elastic and plastic components, respectively. The stress amplitude is indicated by σ_a, and E corresponds to the elastic modulus. The constants K' and n' are material-dependent parameters [30]. It is important to note that these constants differ from those typically associated with the Hollomon equation [31]. The RO equation parameters are provided in Table 1.

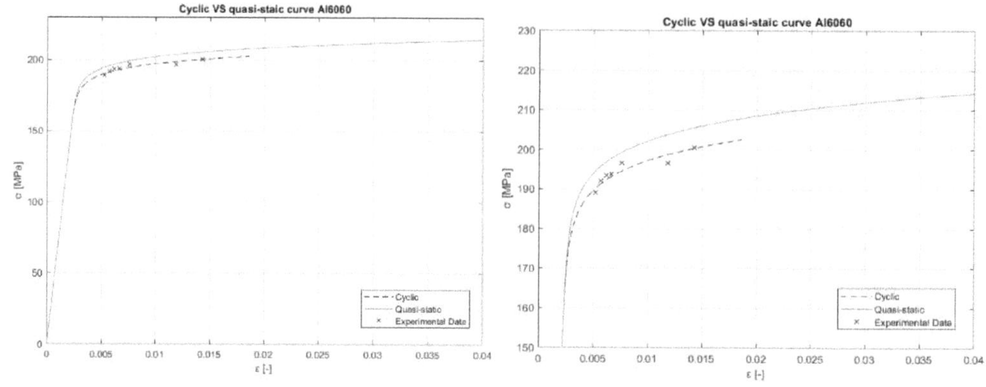

Figure 4: Cyclic vs. quasi-static curve (detail view on the right).

Table 1: RO parameters.

K′	**n′**
233.0807	0.0338

The Al 6060 exhibits softening behaviour. Even after stabilization, testing continued until sample failure to obtain data for fatigue curves, commonly referred to as BCM diagrams (eqn (3)):

$$\varepsilon_a = \frac{\Delta\varepsilon}{2} = \frac{\sigma_f'}{E}(2N)^b + \varepsilon_f'(2N)^c. \tag{3}$$

In this equation, σ_f', b, ε_f' and c represent four material-specific constants that need to be determined, namely the fatigue strength coefficient, the fatigue strength exponent, the fatigue ductility coefficient and the fatigue ductility exponent, respectively. Fig. 5 presents the experimental data obtained, along with the fitted curve using the BCM equation. The calculation process initially involves determining the Basquin coefficients based on the yield stress and high-cycle fatigue limit through linear interpolation. The Basquin equation, which describes the correlation between elastic deformation and the number of cycles, allows for the separation of plastic deformation from the total strain.

$$\varepsilon_{ap} = \varepsilon_a - \varepsilon_{ae} = \varepsilon_f'(2N)^c. \tag{4}$$

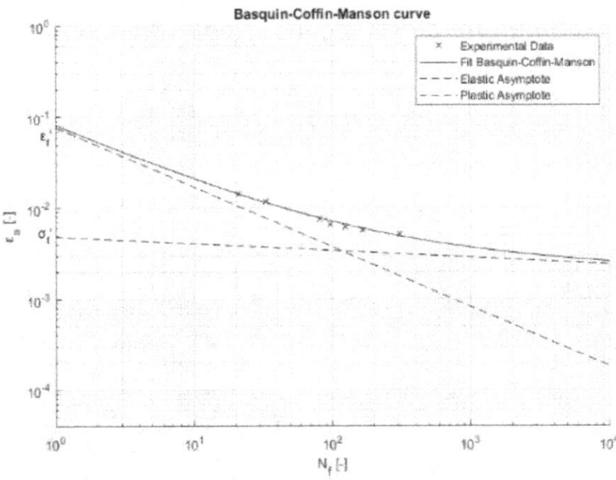

Figure 5: BCM curve.

To estimate the maximum likelihood parameters for b and c in accordance with eqn (4), the ASTM 739 standard [32] was considered. The equation for a best-fit line based on experimental data is expressed as follows:

$$\log(N) = \hat{A} + \hat{B}\log(\varepsilon_{ap}). \tag{5}$$

The maximum likelihood estimators for parameters A and B are given by $\hat{A} = \overline{Y} - \hat{B}\overline{X}$ and $\hat{B} = \frac{\sum(x_i - \overline{X})(y_i - \overline{Y})}{\sum(x_i - \overline{X})^2}$, where $\overline{X}$ and $\overline{Y}$ represent the average values of the independent x_i ($\log(\varepsilon_{ap})$) and independent variable y_i ($\log(N)$).

Eqn (5) can be reformulated as follows:

$$\varepsilon_{ap} = 10^{-\frac{\hat{A}}{\hat{B}}}\left(\frac{1}{2}\right)^{\frac{1}{\hat{B}}}(2N)^{\frac{1}{\hat{B}}}. \tag{6}$$

Thus, the maximum likelihood estimators for the Coffin–Manson coefficients, c and ε_f' can be expressed as:

$$\varepsilon_f' = 10^{-\frac{\hat{A}}{\hat{B}}}\left(\frac{1}{2}\right)^{\frac{1}{\hat{B}}} \tag{7}$$

$$c = 1/\hat{B} \tag{8}$$

The final computed values are provided in Table 2.

Table 2: BCM coefficients.

σ'_f (MPa)	b	ε'_f	c
351.9	−0.076	0.121	−0.657

5 EXPERIMENTAL SETUP

Some samples were cut from the tested specimens to observe the microstructural characteristics. After the standard metallographic preparation by mounting the specimens in a cold-setting resin and by mirror polishing, the samples were observed after the chemical etching (Keller's and Weck's reagents).

Fig. 6 shows the observed microstructure by light optical and scanning electron microscopes (LOM and SEM, respectively). Polygonal grains were clearly visible in both the LOM and SEM images. The average grain size is about 43 μm. The particles remarked in Fig. 6(d) were analysed by EDS probe. They are Fe-rich phases as reported in Table 3.

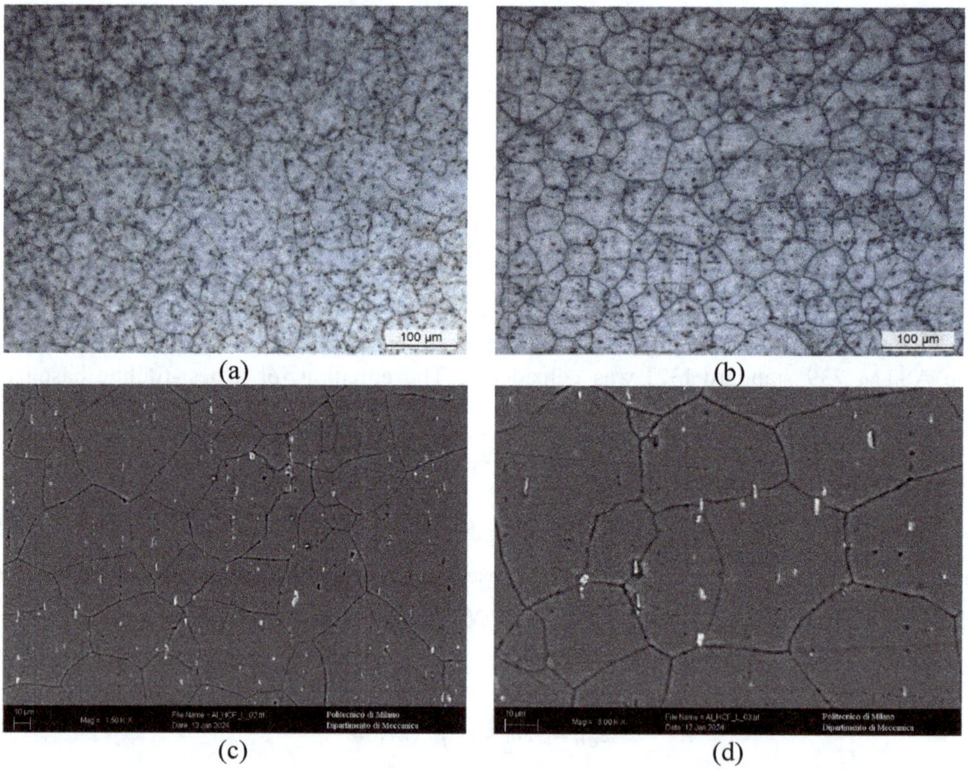

Figure 6: Metallographic analysis of the investigated samples. (a) Transversal HCF section, LOM; (b) Longitudinal HCF section, LOM; (c) Longitudinal HCF section, SEM; (d) Magnification of (c).

Table 3: EDS analyses on the particles remarked in Fig. 7(a).

	Mg (%)	Si (%)	Fe (%)
A	0.65	3.10	6.60
B	0.50	4.63	12.39
C	0.35	3.09	4.36

6 ANALYSIS OF THE FRACTURE SURFACES

Figs 7 and 8 show the fracture surfaces of some of the tested HCF and LCF specimens. In the nucleation zone, an intergranular fracture can be seen. It is an infrequent mechanism that can be related to the precipitation of Mg-Si rich particles at grain boundary. Li et al. [33] reported that such kind of failure can occur in ductile metals only under particular conditions of temperature, strain rate and interstitial atoms content. As confirmed by other authors [34]–[37], it is believed that fatigue cracking can be generated at grain boundaries when persistent slip bands (PSBs) are blocked by grain boundaries creating a stress concentration. Moreover, the presence of intermetallic phases such as Mg-Si precipitates can enhance the formation of the first micro-voids.

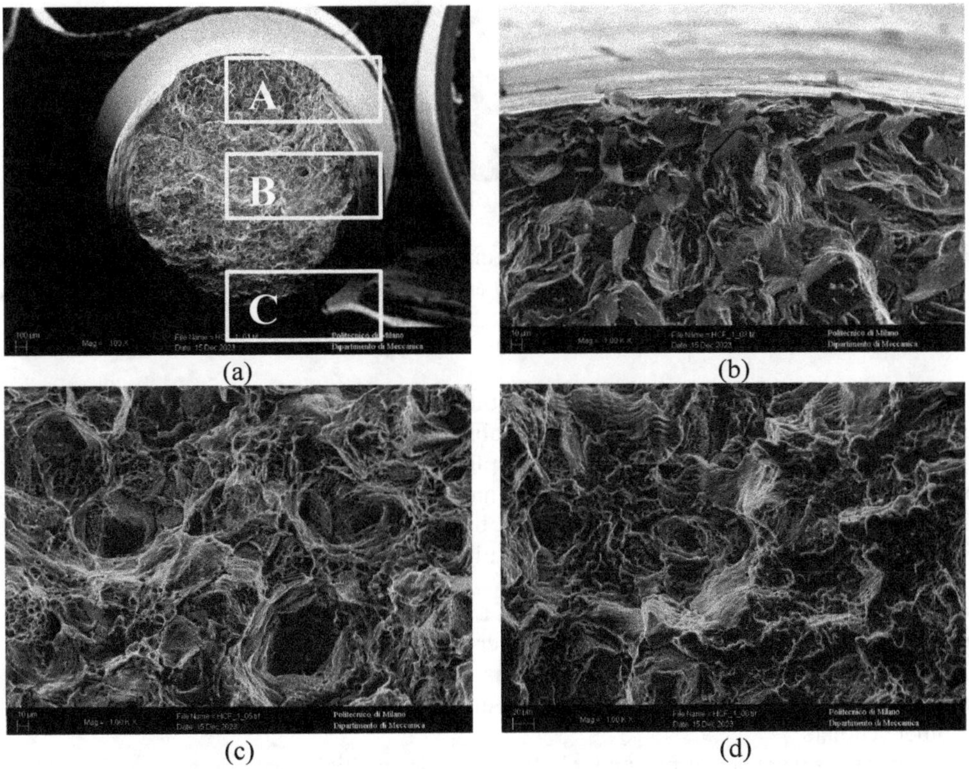

Figure 7: Fracture surface of one HCF specimen. Zone A represents the nucleation zone. The crack propagates downward across zones B and C. Pictures (b), (c) and (d) are magnifications of Zones A, B and C respectively.

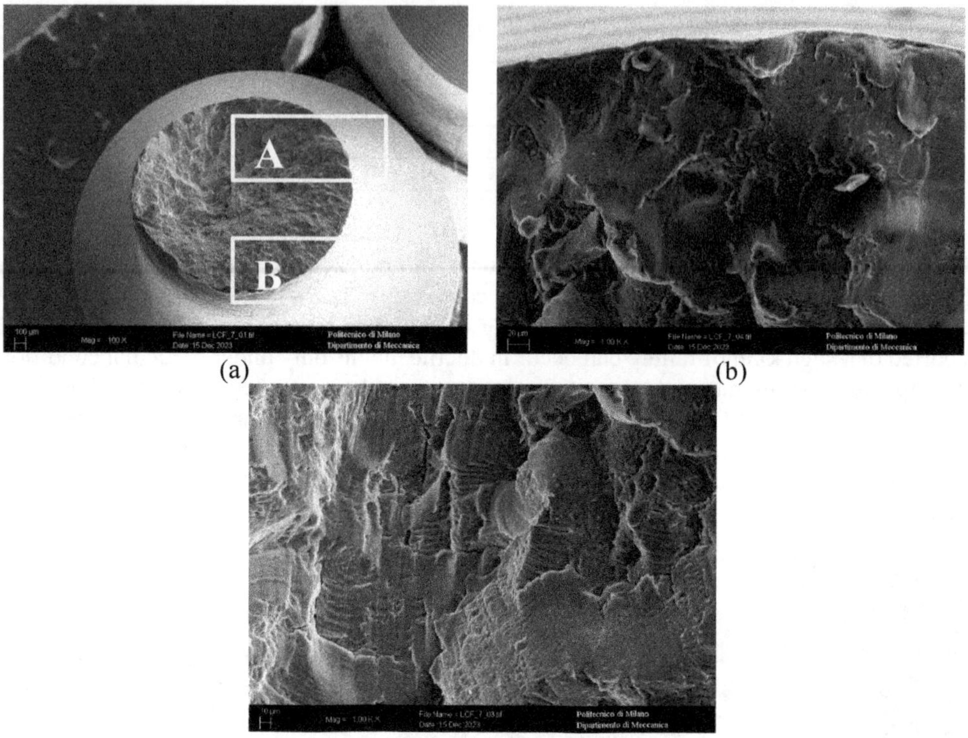

(a) (b) (c)

Figure 8: Fracture surface of one LCF specimen. Zone A represents the nucleation zone. The crack propagates downward across zones B. (b) and (c) are magnifications of Zones A and B, respectively.

The intergranular characteristic is less visible in Zone B and disappear in Zone C. In zones B and C dimples and some striations can be observed as well.

The LCF fracture surfaces are more complicated to be studied since some damages are present probably because of the higher loads and of the load ratio ($R = -1$). No intergranular feature can be seen, but it can be hidden by the local deformations visible in Fig. 8. Nevertheless, striations are clearly visible in Fig. 8(c).

7 CONCLUSIONS

In this work an Al 6060 alloy was tested under static and dynamic loadings.

The quasi-static test allows the determination of the elastic modulus and the ultimate tensile stress (69 GPa and 219 MPa, respectively). These values are aligned with the findings of other scholars.

Fatigue tests were performed under both low-cycle and high-cycle-fatigue conditions. The cyclic behaviour was modelled via the RO model showing a marked softening behaviour. The high-cycle-fatigue limit was determined exploiting the short stair-case approach proposed by Dixon, resulting in 143.8 MPa. The strain–cycle curve was interpolated using the BCM model. The results are aligned with the finding of other scholars.

The peculiarity of this research is in the observed fracture mechanism. In particular, an intergranular fracture was observed. It is an occasional mechanism related to the precipitation of Mg-Si rich particles at grain boundary that can occur in ductile metals only under particular conditions of temperature, strain rate and interstitial atoms content.

REFERENCES

[1] Kaufman, J.G. & Holt, M., Fracture characteristics of aluminum alloys. Technical paper no. 18, Aluminum Company of America, Pittsburgh, PA, 48 pp., 1965.

[2] Kaufman, J.G., Aluminum alloys for Arctic applications. *Proceedings of the Conference on Materials Engineering in the Arctic*, St. Jovite, Quebec, 1976.

[3] Bucci, R.J., Selecting aluminum alloys to resist failure by fracture mechanisms. *Eng. Fract. Mech.*, **12**, pp. 407–441, 1979.

[4] Wagener, R. & Melz, T., Fatigue life curve: A continuous Wöhler curve from LCF to VHCF: Dedicated to Professor Dr. Ing. Harald Zenner on the occasion of his eightieth birthday. *Materials Testing*, **60**, pp. 924–930, 2018.

[5] Pisapia, A., Nastri, E., Piluso, V., Formisano, A. & Mazzolani, F.M., Experimental campaign on structural aluminium alloys under monotonic and cyclic loading. *Eng. Struct.*, **282**, 115836, 2023.

[6] Hopperstad, O.S., Langseth, M. & Remseth, S., Cyclic stress–strain behaviour of alloy AA6060, Part I: Uniaxial experiments and modelling. *Int. J. Plast.*, **11**, pp. 725–739, 1995.

[7] Jha, S.K., Balakumar, D. & Paluchamy, R., Experimental analysis of mechanical properties on AA6060 and 6061 aluminum alloys. *Int. J. Eng. Res. Appl.*, **5**, pp. 47–53, 2015.

[8] Fernandes, F.A.O., Gonçalves, J.J.M. & Pereira, A.B., Evaluation of laser lap weldability between the titanium alloy Ti-6Al-4V and aluminum alloy 6060-T6. *Crystals (Basel)*, **13**, p. 1448, 2023.

[9] Hockauf, K., Meyer, L.W., Halle, T. & Hockauf, M., Rissfortschritts-und ermüdungsverhalten der aluminiumlegierung EN AW-6060 nach ECAP und nachgelagerter wärmebehandlung. *Materialwissenschaft und Werkstofftechnik: Entwicklung, Fertigung, Prüfung, Eigenschaften und Anwendungen Technischer Werkstoffe*, **40**, pp. 493–499, 2009.

[10] Hockauf, K., Niendorf, T., Wagner, S., Halle, T. & Meyer, L.W., Cyclic behavior and microstructural stability of ultrafine-grained AA6060 under strain-controlled fatigue. *Procedia Eng.*, **2**, pp. 2199–2208, 2010.

[11] Borrego, L.P., Abreu, L.M., Costa, J.M. & Ferreira, J.M., Analysis of low cycle fatigue in AlMgSi aluminium alloys. *Eng. Fail. Anal.*, **11**, pp. 715–725, 2004.

[12] Corre, T., Perrin, J., Gebhard, J., Duchateau, T., Lilensten, L., Laurent-Brocq, M., Tekkaya, A.E. & Huneau, B., Tensile and shear behavior of recycled AA 6060 aluminium chips by direct hot extrusion. *Materials Science and Technology*, 02670836241287768, 2024.

[13] Pardoen, T., Dumont, D., Deschamps, A. & Brechet, Y., Grain boundary versus transgranular ductile failure. *J. Mech. Phys. Solids*, **51**, pp. 637–665, 2003.

[14] Prince, K.C. & Martin, J.W., The effects of dispersoids upon the micromechanisms of crack propagation in Al Mg Si alloys. *Acta Metallurgica*, **27**, pp. 1401–1408, 1979.

[15] Christ, H.-J. & Mughrabi, H., Cyclic stress–strain response and microstructure under variable amplitude loading. *Fatigue Fract. Eng. Mater. Struct.*, **19**, pp. 335–348, 1996.

[16] Kowalski, A., Ozgowicz, W., Jurczak, W., Grajcar, A., Boczkal, S. & Kurek, A., Microstructural and fractographic analysis of plastically deformed Al-Zn-Mg alloy subjected to combined high-cycle bending-torsion fatigue. *Metals (Basel)*, **8**, p. 487, 2018.

[17] Kowalski, A., Ozgowicz, W., Grajcar, A., Lech-Grega, M. & Kurek, A., Microstructure and fatigue properties of AlZn6Mg0. 8Zr alloy subjected to low-temperature thermomechanical processing. *Metals (Basel)*, **7**, p. 448, 2017.

[18] ASTM, Standard Practice for Conducting Force Controlled Constant Amplitude Axial Fatigue Tests of Metallic Materials. ASTM, 2007.

[19] Concli, F. & Gilioli, A., Numerical and experimental assessment of the static behavior of 3D printed reticular Al structures produced by selective laser melting: progressive damage and failure. *Proceedings of the Procedia Structural Integrity*, **12**, pp. 204–212, 2018.

[20] Concli, F., Gilioli, A. & Nalli, F., Experimental–numerical assessment of ductile failure of additive manufacturing selective laser melting reticular structures made of Al A357. *Proc. Inst. Mech. Eng. C. J. Mech. Eng. Sci.*, **235**, pp. 1909–1916, 2021.

[21] Nalli, F., Cortese, L. & Concli, F., Ductile damage assessment of Ti6Al4V, 17-4PH and AlSi10Mg for additive manufacturing. *Eng. Fract. Mech.*, **241**, 2021.

[22] Alomar, Z. & Concli, F., Compressive behavior assessment of a newly developed circular cell-based lattice structure. *Mater. Des.*, **205**, 2021.

[23] Concli, F. & Gilioli, A., Numerical and experimental assessment of the mechanical properties of 3D printed 18-Ni300 steel trabecular structures produced by selective laser melting: A lean design approach. *Virtual Phys. Prototyp.*, **14**, pp. 267–276, 2019.

[24] Dixon, W.J., The up-and-down method for small samples. *J. Am. Stat. Assoc.*, **60**, pp. 967–978, 1965.

[25] Gorla, C. et al., Bending fatigue strength of case carburized and nitrided gear steels for aeronautical applications. *International Journal of Applied Engineering Research*, **12**, pp. 11306–11322, 2017.

[26] Concli, F., Austempered ductile iron (ADI) for gears: Contact and bending fatigue behavior. *Procedia Structural Integrity*, **8**, pp. 14–23, 2018.

[27] Gorla, C., Conrado, E., Rosa, F. & Concli, F., Contact and bending fatigue behaviour of austempered ductile iron gears. *Proc. Inst. Mech. Eng. C. J. Mech. Eng. Sci.*, **232**, pp. 998–1008, 2018.

[28] Gorla, C., Rosa, F., Concli, F. & Albertini, H., Bending fatigue strength of innovative gear materials for wind turbines gearboxes: Effect of surface coatings. *Proceedings of the ASME International Mechanical Engineering Congress and Exposition*, **45233**, pp. 3141–3147, 2012.

[29] ASTM, Standard Test Method for Strain-Controlled Fatigue Testing. ASTM E606/E606M, ASTM International, 2012.

[30] Ramberg, W. & Osgood, W.R., *Description of Stress–Strain Curves by Three Parameters*, 1943.

[31] Hollomon, J.H., Tensile deformation. *Aime Trans*, **12**, pp. 1–22, 1945.

[32] ASTM, Standard Practice for Statistical Analysis of Linear or Linearized Stress-Life (s–n) and Strain-Life (e–n) Fatigue Data. ASTM 739-91, 2006.

[33] Li, Y., Qiu, S., Zhu, Z., Han, D., Chen, J. & Chen, H., Intergranular crack during fatigue in Al-Mg-Si aluminum alloy thin extrusions. *Int. J. Fatigue*, **100**, pp. 105–112, 2017.

[34] Islam, M.N. & Arai, Y., Ultrasonic back reflection evaluation of crack growth from PSBs in low-cycle fatigue of stainless steel under constant load amplitude. *Materials Science and Engineering: A*, **520**, pp. 49–55, 2009.

[35] Essmann, U., Gösele, U. & Mughrabi, H., A model of extrusions and intrusions in fatigued metals I. Point-defect production and the growth of extrusions. *Philosophical Magazine A*, **44**, pp. 405–426, 1981.

[36] Taheri, S., Hauet, A., Taleb, L. & Kpodekon, C., Micro–macro investigations about the fatigue behavior of pre-hardened 304L steel. *Int. J. Plast.*, **27**, pp. 1981–2004, 2011.
[37] James, M.N., Intergranular crack paths during fatigue in interstitial-free steels. *Eng. Fract. Mech.*, **77**, pp. 1998–2007, 2010.

SECTION 3
EXPERIMENTS AND NUMERICAL ANALYSIS

EXPERIMENTAL TESTS AND FINITE ELEMENT ANALYSES OF THE DEMOUNTABLE STEEL TRUSS JOINTS WITH RECTANGULAR HOLLOW SECTIONS

ABDULLAH AL-MAISARI, NEWTON AVILA DOS SANTOS, MIKI SALLI & ZHONGCHENG MA
Häme University of Applied Sciences, HAMK Tech Research Unit, Finland

ABSTRACT

An innovative demountable solution for the steel roof trusses with rectangular hollow sections (RHS) was developed to promote the reuse of steel roof trusses. Self-drilling screws were used for the RHS truss joints to replace the traditional welded solution. The drillability of screwed connections for S355 and S500 steel plates was tested in terms of drilling time and failed attempts. The effects of the size of pilot holes and the type of screws were studied. Lap-shear behaviour and strength with varying gap distances between the plates were also studied experimentally. Finally, three T-joint tests in tension were executed to investigate the load-carrying capacity of the joints. Digital image correlation technique was used to record the deformation of joint area. Finite element analyses were firstly performed for the lap-shear tests using simplified modelling methods. Then, the simplified modelling method of the screwed connection was validated by the T-joint tests.

Keywords: demountable joint, RHS, self-drilling screws, drillability, lap-shear behaviour, T-joint.

1 INTRODUCTION

The roof truss structure with rectangular hollow sections (RHS) has been widely used in Nordic countries, especially in Finland. The tubular chords and braces are welded together to form the structures. The design of truss joints with welded RHS is well specified by EN 1993-1-8.

In recent years, the demountable solutions for RHS truss joints attract the research and innovations. Non-weld truss joints are very common in cold-formed steel truss structures with open sections. Screw connection, bolt connection, press-joining connection, and rosette connection are typical applications [1]–[3]. However, the demountable solutions of non-weld RHS truss joint is very challenging since the inner surface of RHS is inaccessible. With the advent of laser cutting technology, the complex shape of cutting can be easily made, the truss joint can be made by plug and play contact without welding [4]–[6].

The use of modern 3D laser cutting technology enables easy and precise cutting in steel elements including enclosed RHS. The shape of the cutting can be very complex and easily controlled. In the developed non-welded RHS truss joints by Szlendak and Szpyrka for tension braces [4], forces are transferred only by bearing and shearing stresses in the elements in contact with each other, the key (bracing elements) and the lock (holes in the chord). Due to lack of permanent connections between the elements, such structure can be dismantled and reassembled elsewhere.

A demountable joint for tubular sections using tab–slot configuration without welds and screws was patented by Iriarte [7]. The cuts and slots can be made by laser cutting or plasma cutting. However, the perceived strength of the joint for building truss application may be too weak to resist the tensile and compression forces in the braces.

In this paper, a novel demountable solution for RHS trusses using self-drilling screws and computer numerical control (CNC) cutting process is developed and the drillability and mechanical behaviour are investigated both experimentally and numerically. The mechanical tests include lap-shear tests and T-joint tests.

WIT Transactions on The Built Environment, Vol 216, © 2025 WIT Press
www.witpress.com, ISSN 1743-3509 (on-line)
doi:10.2495/HPSM250051

2 DRILLABILITY TESTS

Drillability of self-drilling screws refers to the ability of a screw to penetrate and fasten into a material efficiently or without failure. It is influenced by factors such as: the diameters of pre-drilled holes, the length of the drilling bits of screws, plate material hardness, and rotational speed and torque. The purpose of drillability tests is to find out the influence of these parameters on the drilling times and successful attempts. Prior to this drillability study, the truss joints with pilot holes were cut using laser cutting. However, nearly all drilling attempts failed due to the laser hardening effect along the edges of round holes. To address this issue and further investigate the drillability, pilot holes were mechanically drilled instead of using laser cutting in this study.

Self-drilling screws are typically used to fasten a thin sheet (max. 2 mm) made from S280GD up to S550GD to a thicker substructure made of structural steel grade up to S355. Drilling capacities of these screws increase as the screw diameter increases but is usually less than 6 mm. The long drill bit versions might be able to drill through 13.5 mm thick plates. The European Technical Assessment (ETA) for these screws usually only include design situations in which thin sheets are being attached to thicker substrates. Therefore, it is important to find out how these screws perform with thicker plates that might also be made from high-strength steel. The effect of small gaps between plates on the drillability needed to be investigated as well. Moreover, the quantitative effects of the diameters of pilot holes on the drilling times and successful attempts need to be further studied.

2.1 Test method and cases

The test objective is to investigate how self-drilling screws with ordinary and long drill bits work with 4 mm and 8 mm thick steel plates made from S355 and S500 when there is a small 1–2 mm gap between the plates to be connected. The test variables that affect the outcome of the tests are pilot hole size, drilling speed, plate thicknesses, plate material, screw size, screw type, gap size, and drilling maximum torque. The screw options for the tests are shown in Fig. 1.

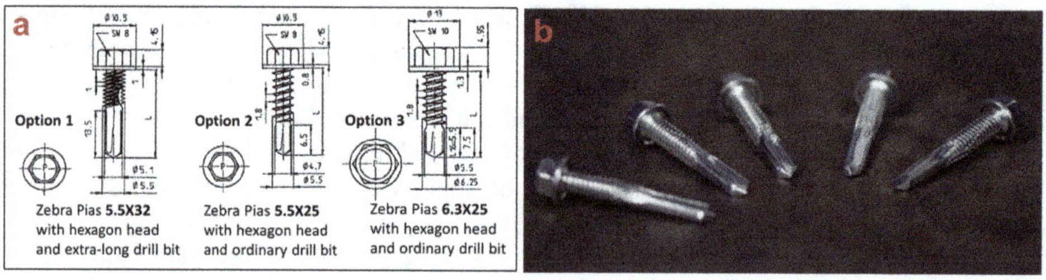

Figure 1: (a) Screw options; and (b) PIAS 5.5 × 32 with a long drill bit.

Testing has been implemented by manufacturing the test specimens which consist of a pair of plate pieces (Fig. 2) that have been fixed together utilizing small bolts. In test cases with a gap between the plates, 1 mm thick washers have been added to produce the desired gaps between two steel plates. Pilot holes of varying size have been pre-drilled to the plates. Pilot hole sizes 3.5 mm, 4.0 mm and 4.5 mm as well as the case without pilot holes have been investigated with three different screw options. Plates were fixed by a bench vice, and screws were driven to the plates using a battery-operated drill. Drilling times were recorded in six attempts in each case.

Figure 2: Example of (a) test specimen with screws drilled in; and (b) drillability test assembly pair of plates (4 + 8 mm plates with 1 mm gap).

There are 144 different test cases. To investigate the torque and the drilling time, each case is repeated nine times. This means driving 1,296 screws into the plates. In addition to these tests, also drilling speed (measured in revolutions per minute (RPM)) effect has been studied to select a suitable RPM for the tests. The preliminary study for the drilling speed has been done for S355 and S500, 8 + 4 mm plates without pilot holes to demonstrate an extreme situation.

2.2 Test results

RPM tests showed that the optimal drilling speed is 1400 rpm for 8 mm thick S500 plates and 1700 rpm for S355 plates. High speed combined with moderate push force and a high strength steel (HSS) substrate produce excessive heat which limits drill bit endurance in thicker HSS material. Drillability performance was improved with thinner plates and pre-made pilot holes. The Milwaukee ONEDD3 drill was used, with torque settings up to 158 (Nm).

Tests on screw options 1, 2, and 3 (as shown in Fig. 1(a)) revealed that screws with longer drill bits performed well for both S355 and S500 plates. However, for thicker 8 mm plates, especially without properly sized pilot holes, a reduced drilling speed was necessary for success. It was found that screws with standard drill bits frequently failed due to the exceedance of their drilling capacity. Pre-made pilot holes significantly reduced drilling time and improved success rates. Figs 3 and 4 present the average drilling times and failure rates for PIAS 5.5 × 32 screws with long drill bits in S355 and S500 plates, respectively.

3 LAP SHEAR AND T-JOINT TESTS

Force transfer between the brace members and the chord member in a mechanical truss joint is based on a shear force transfer mechanism. Joint forces are transferred between the plate walls of the tubular profiles by the screw connectors. This behaviour can be mimicked in simple lap shear tests. Tests are carried out to obtain information on the strength and deformation of axially loaded single lap connections of steel plates with self-drilling screws and a possibility of a small gap between the connected plates.

3.1 Description of lap-shear tests

Test assemblies for the lap shear tests (Fig. 5) are composed of pairs of plates with a size of 60 mm × 300 mm. The thicknesses of the plates are 4 mm and 8 mm, respectively and made

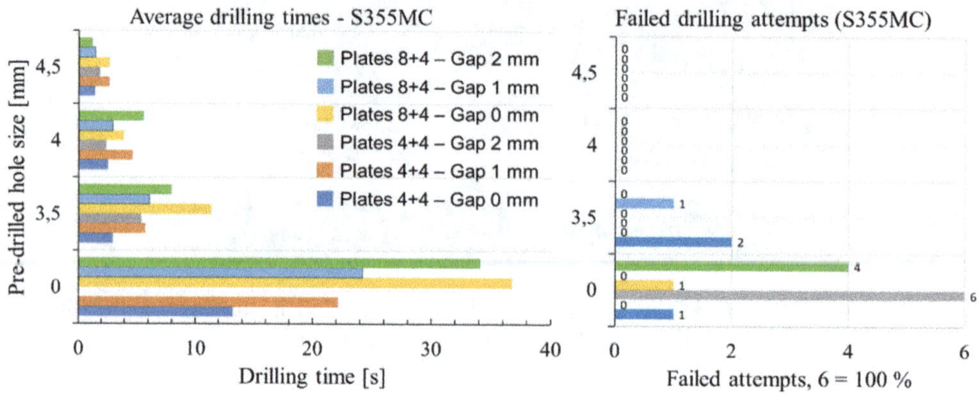

Figure 3: Drillability test results for PIAS 5.5 × 32 screws in S355 plates.

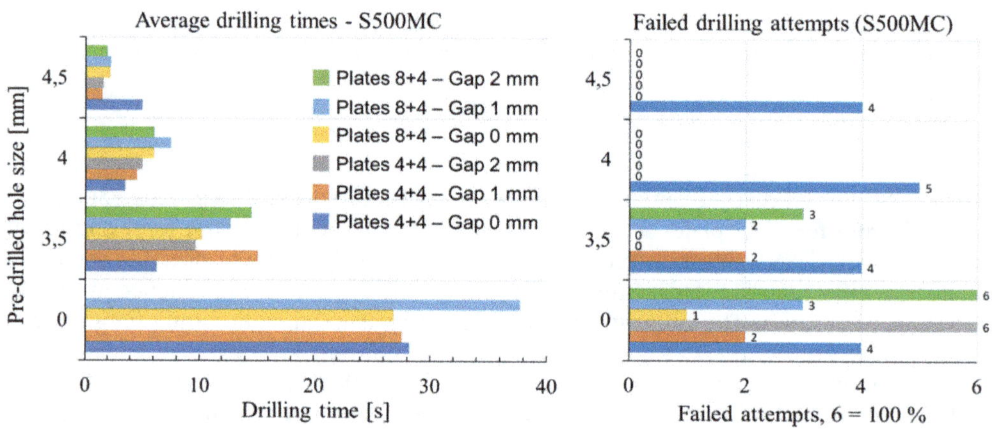

Figure 4: Drillability test results for PIAS 5.5 × 32 screws in S500 plates.

from steel grades S355MC and S500MC. To eliminate gripping problems with the 50 kN tensile testing machine, the ends of the test plates are grinded to bare steel. Four holes for M4 fixing bolts are drilled before the self-drilling screw is added. The bolts are removed once the test assembly has been gripped to the tensile testing machine.

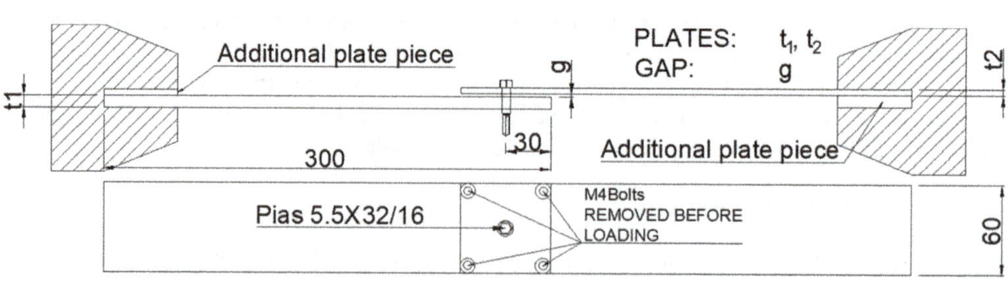

Figure 5: Lap shear test assembly.

End distance (e_1) and edge distance (e_2) of the connection are chosen so that they do not reduce the bearing strength of the connection. For self-drilling screws, the nominal diameter of the screw can be used for a conservative approximation. Gaps are created using 1 mm washers, and additional (filler) plates in the ends are added to eliminate setup eccentricity as shown in Fig. 5.

Screw option 1 (ZEBRA Pias 5.5 × 32) was selected for the lap shear tests due to its strong performance in drillability tests. As shown in Table 1, a total of 36 tests were conducted across 12 test series, with each series consisting of three individual tests. The tests were performed on single fasteners while varying key parameters: plate thickness (4 + 4 mm/ 8 + 4 mm), plate gap (0 mm/1 mm/2 mm), and steel grade (S355MC/S500MC). Testing was displacement-controlled at 0.2 mm/min, using extensometer sensor arms for strain measurement and a load cell for force measurement. The extensometer gauge length was set to 100 mm.

Table 1: Lap shear test results.

Lap shear test cases	Steel grade	Average shear strength (kN)
4 + 4 mm Plates, Gap = 0 mm	S 355 MC	10.93
8 + 4 mm Plates, Gap = 0 mm		10.20
4 + 4 mm Plates, Gap = 1 mm		9.34
8 + 4 mm Plates, Gap = 1 mm		8.92
4 + 4 mm Plates, Gap = 2 mm		8.31
8 + 4 mm Plates, Gap = 2 mm		8.04
4 + 4 mm Plates, Gap = 0 mm	S 500 MC	10.62
8 + 4 mm Plates, Gap = 0 mm		10.70
4 + 4 mm Plates, Gap = 1 mm		9.69
8 + 4 mm Plates, Gap = 1 mm		9.16
4 + 4 mm Plates, Gap = 2 mm		8.81
8 + 4 mm Plates, Gap = 2 mm		8.31

3.2 Results of lap-shear tests

Table 1 lists the measured average shear strength for lab-shear tests for different steel grades, plate thicknesses and gap dimensions. The pre-drilled hole in all the tests was 4.0 mm in diameter and the screw type was Pias 5.5 × 32. The average shear strength for screws was the average out of three successful tests.

As shown in Table 1 and Fig. 6, the gap effect on the shear strength of the screw is quite considerable in both steel grades S355MC and S500MC. The shear strength of 5.5 × 32 screw dropped about 30% in the case of S355MC and 20% in the case of S500 MC steel plates with the increasing gap distances from zero to 2 mm. The average shear strength is not affected significantly between grades S355MC and S500MC plates both with and without a gap. The elastic stiffness of the lap-shear joint for S500 plates is higher than S355 plates without a gap while remains similar for those with 2 mm gap. Ductility is slightly higher in the case of 2 mm gap compared to zero gap between plates. However, the overall behaviour of lap-shear joints demonstrates brittle failures.

For the 8 + 4 mm plate case with S355 steel (Fig. 7), the ultimate shear strength decreased by approximately 7% compared to the 4 + 4 mm plates for the cases with zero gap. This reduction may be attributed to the larger engagement area (12 mm) between the screw thread

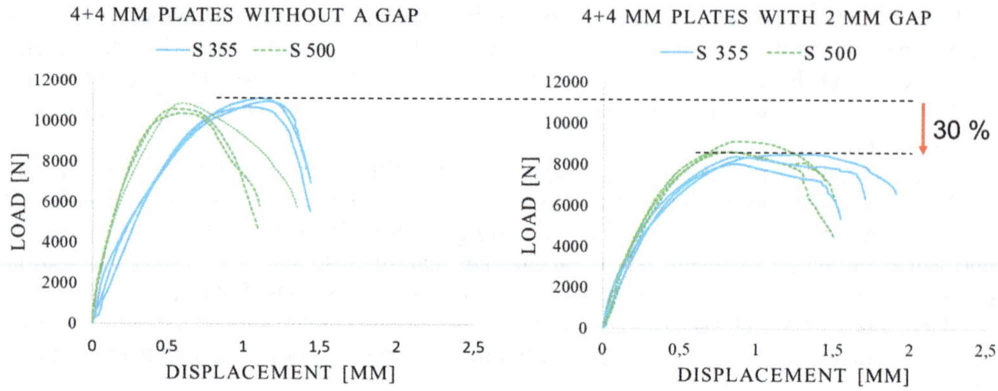

Figure 6: Comparison of lap shear test results for 4 + 4 mm plates with and without gaps for different steel grades.

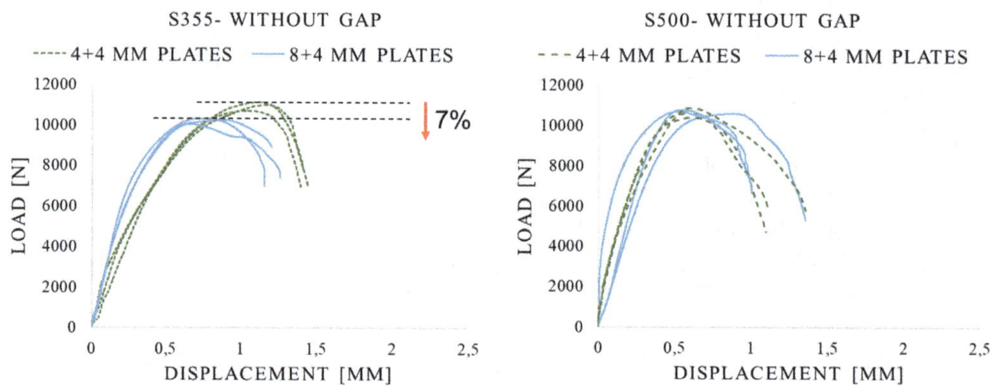

Figure 7: Comparison of lap shear test results for different steel grades and plate thickness arrangements without gaps.

and the plates. However, in the case of S500 steel, the shear strength for the 8 + 4 mm plates remained nearly the same as for the 4 + 4 mm plates. This was the only instance where the 4 + 8 mm plate arrangement did not negatively impact shear strength. In all other test configurations, the 4 + 8 mm plates resulted in a shear strength reduction of 3.3% to 7.15% depending on the gap distances.

In these tests, the plate arrangement consistently followed the sequence of the 8 mm plate first, followed by the 4 mm plate. The 8 mm plates were drilled first, as the thicker chord members are typically drilled before the braces in the construction of steel trusses for this type of demountable mechanical joint solution.

As shown in Fig. 8, the primary failure mode observed was shear failure of the screw, with the fracture plane aligning parallel to the sliding surfaces. As expected, the plates remained less deformed due to their higher bearing strength. Prior to failure, the screw exhibited gradual slight tilting under load before ultimately shearing failure.

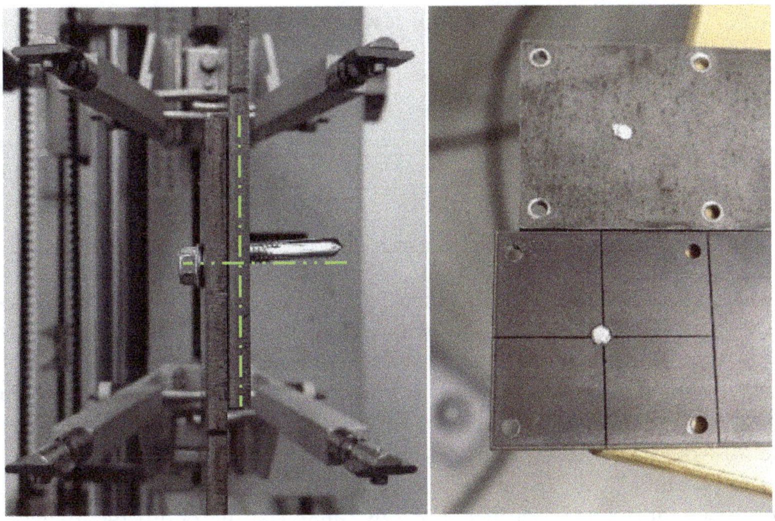

Figure 8: Shear failure of screw – lap-shear test.

3.3 Description of T-joint tests

The purpose of the T-joint tests in this study was to evaluate the influence of screw spacing distances on the stiffness and load-bearing resistance of demountable mechanical joints under tension load. The shear strength of the demountable T-joint solutions was derived from lap shear test results. Fig. 9 shows the test specimen and setup for T-joint test in tension. The cases reported in this paper are part of a more comprehensive test program which will be carried out shortly later.

For specimen assembly, steel tubes are cut to a length of 600 mm. A total of 24 predrilled 4.0 mm holes are created, followed by the installation of 12 Pias 5.5 × 32 screws on each side of the chord. Table 2 shows the details of the T-joint test cases.

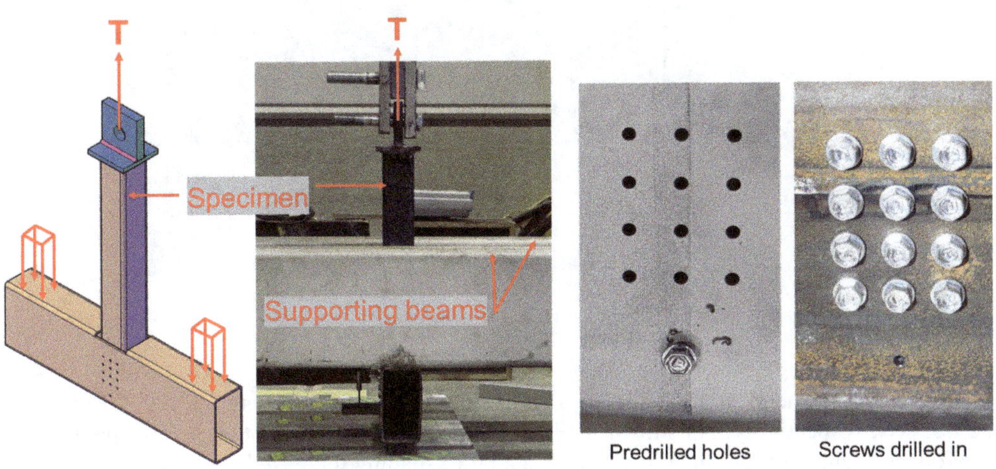

Figure 9: Test specimen and setup for T-joint tests.

Table 2: T-joint test cases in tension.

Case	Chord section	Brace section	Screws spacing (mm)	No. of screws per side	Steel grade	Predrilled hole size (mm)
4	RHS 160X80X4	SHS 70X70X4	15	12	S420	4.0
5			20			
6			25			

The end distance (e_1) and edge distance (e_2) of the connection are done according to EN 1993-1-8, $e_1 \geq 3d_0$ and $e_2 \geq 1.5d_0$, where d_0 is the screw hole diameter. The number of screws is determined based on the estimated shear resistance of the screws.

The tests were displacement-controlled at a loading rate of 0.8 mm/min, utilizing a tensile frame with a capacity of 250 kN. Digital image correlation (DIC) was employed to precisely capture the full-field strain distributions across the joint. In addition, a linear variable differential transformer (LVDT) sensor was installed to measure the relative displacements between the brace and chord member. For DIC analysis, the chord member with the gap between the chord and brace (identified as the weaker component) was speckle-painted to facilitate DIC strain tracking.

3.4 Results of T-joint tests

As illustrated in Fig. 10, the screws exhibited pure shear failure under loading. Minor bending of screw shanks was observed in some cases. Additionally, localized deformations occurred at the chord holes due to bearing stresses from the screws. Variations in hole bearing deformation were observed after removing the failed screws and attributed to the misalignment between brace and chord holes due to the manufacturing tolerances and drilling issues.

Figure 10: Failure modes.

The ultimate shear strength for T-joint test cases 4, 5 and 6 were 217.02, 225.09 and 229.82 kN, respectively. As shown in Fig. 11, the results revealed that screw spacing has obvious influences on the shear resistance: a 25 mm spacing increased average shear strength by 6% compared to a 15 mm spacing. Moreover, ductility increased by about 30% in case 6 with 25 mm spacing compared to case 4 with only 15 mm spacing.

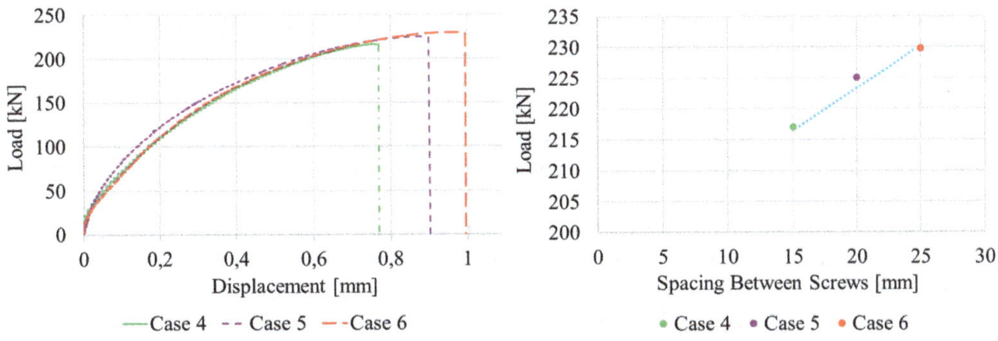

Figure 11: (Left) Load versus displacement curves for T-joint tests; and (Right) Effect of spacing on shear strength of T-joints.

Additionally, it was found that the average shear strength of single screws in the group decreases compared to lap-shear test results. The reduction is probably due to the rigidity of the joint since rotation of screws is prevented and the gap between brace and chord walls in one side. While some screws were drilled twice during assembly, it may have minor effect on joint performance. However, minor hole misalignment caused by machining tolerances led to slightly uneven bearing resistance across holes, which may also explain some challenges during assembly.

The use of LVDTs proved inadequate for capturing precisely the relative displacements between the brace and chord. To improve the instrumentation of the tests, strain gages will be implemented in future tests to validate the DIC measurement.

4 FE ANALYSYS

4.1 FE models

Multiple finite element (FE) analyses were performed using ABAQUS/Explicit software and the models were validated by the lap shear and the T-joint experimental tests. Nonlinear geometric and material behaviours were incorporated into the numerical models. The steel plates of both lap shear and T-joint tests were modelled with S420 steel grade and elasto-plastic material properties obtained from the literature [8]. The yield strength, ultimate stress, and Young's modulus applied to the numerical models were 518.46 MPa, 567.93 MPa, and 210 000 MPa, respectively. The mechanical properties, originally represented as engineering stress–strain curves, were converted into trues stress and effective plastic strain curves.

The lap shear experiments were carried out on S355 and S500 steel plates whereas S420 tubular sections were utilized for all T-joint experiments. Therefore, the lap shear experimental results of S355 and S500 plates were interpolated to obtain the results of S420 plates and further used for the FE models' validation.

ABAQUS connector elements were used to model the screws. Three-dimensional eight-node solid hexahedral elements (C3D8) were employed to model the full geometry of the steel plates in the numerical analyses for both lap shear joints and T-joints. A refined mesh was used around the screw holes to capture local stress concentrations, while a coarser mesh was employed for the areas away from these critical areas aiming to optimize computational efficiency. Boundary conditions in the numerical analyses mirrored the experimental test setups. In these explicit analyses, general interactions including a friction coefficient of 0.1 were defined between contacting steel sheets.

4.1.1 Lap shear tests validation

The lap shear experimental tests were validated in two phases. On the first phase a connector alone was validated whereas on the second phase the lap shear tests including the plates geometry was validated.

A bushing connector, which allows the definition of both translational and rotational components of relative motion, was selected to numerically represent the behaviour of the screws. The end nodes of the connectors were coupled to the nodes on the respective walls of the screw holes. The interpolated S420 lap shear test results were utilized as input to define the nonlinear response of the connectors. The behaviour of the connectors included elastic stiffness, plastic deformation, damage initiation and evolution criteria modelled according to Abaqus manual.

Translational properties derived from the interpolated S420 lap shear tests were applied to the connector in directions transverse to the hole axis of symmetry. Along the screw axis, stiffness properties derived from a Young's modulus of 210,000 MPa and plastic properties based on the screw tensile strength specified by the manufacturer were applied. No restriction was imposed on rotation around the hole's symmetry axis itself. Fig. 12(a) illustrates a bushing connector and Fig. 12(b) shows the comparison between force–displacement curves obtained from the interpolated S420_G0_4+4 lap shear test and the corresponding numerical model consisting only of a bushing connector. The numerical model accurately predicts both force and displacement measurements from the lap shear experiments. Similar accuracy was obtained in all other numerical models employing only bushing connectors for lap-shear tests.

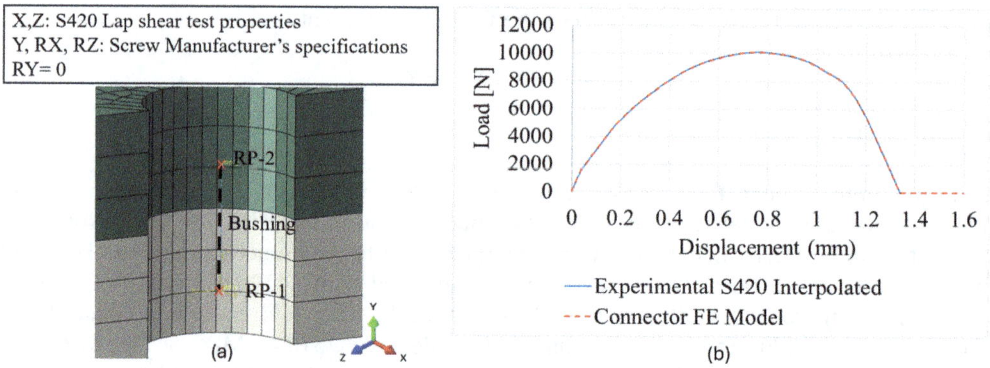

Figure 12: (a) Bushing connector; and (b) Connector and lap shear test force–displacement comparison.

For the validation of the complete lap shear tests, full geometry of the plates was modelled, and the screws were replaced by validated bushing connectors. Connector endpoints were attached to two distinct reference points located at the mid-thickness of the plates, on the centre of the screw holes (Fig. 12(a)). A forced displacement of 2 mm was applied to the end of one plate, while one end of the opposite plate was fixed, replicating experimental test setup. The original results obtained from numerical models indicated good agreement in ultimate load predictions but exhibited slightly lower stiffness compared to experimental results. This stiffness discrepancy likely arose from the neglected steel plate deformations and the simplifications introduced by replacing the screw with a connector.

To reduce these differences and better capture only the screw behaviour in the connector model, the stiffness of the connectors was refined through a reverse engineering approach in the lap shear model. Data points from FE models and experimental lap shear tests were collected at same load levels and the difference in displacements was computed. The additional displacements were subtracted from the connector model. The refined properties for the bushing connector were then applied for the lap shear models. Fig. 13(a) shows the lap shear FE model, while Fig. 13(b) compares experimental and numerical force–displacement curves for the S420_G0_4+4 test. The results indicate that the FE models can accurately capture the ultimate loads and stiffness in the lap shear model. Similar results were obtained in all other lap shear FE analyses. The derived connector properties will be used in the following T-joint FE analyses.

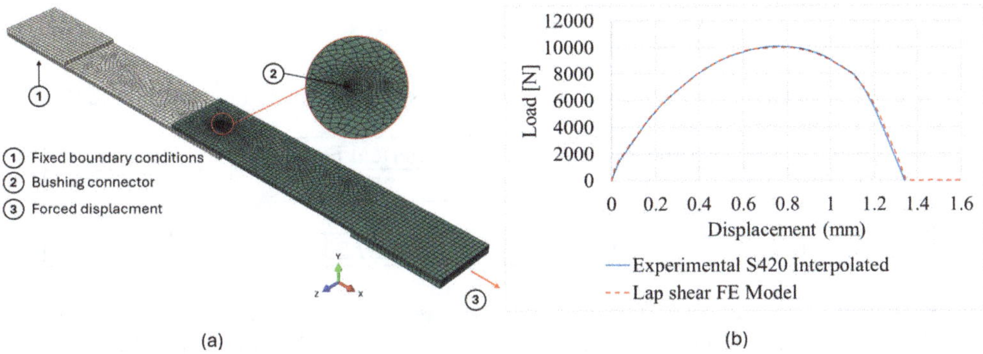

Figure 13: Lap shear. (a) Numerical model; and (b) Lap shear FE and experimental test force–displacement comparison.

4.1.2 T-joint validation

For numerical modelling of T-joints, the full geometry of the chord was modelled, while the vertical brace was represented with a shortened geometry to improve computational efficiency. The screws were modelled by the bushing connectors with properties validated by lap-shear tests. Fig. 14(a) shows the FE model for T-joint Case 06, and Fig. 14(b) presents experimental versus numerical force–displacement results. The numerical models demonstrated slightly lower initial stiffnesses and higher ductility in comparison with the experiments. It is important to note that, for the validation of T-joint FE models, interpolated connector properties for S420 plates were used instead of actual test data, which may have contributed to the observed discrepancy.

Table 3 summarizes the ultimate forces from experimental tests and numerical simulations, including the percentage differences. The numerical models were not able to accurately capture the spacing effect on ultimate load discussed in Section 3.

5 DISCUSSIONS AND CONCLUSIONS

The experimental part of this study involved three tests: drillability, lap-shear, and T-joint tests. Over 144 drillability tests showed that the optimal drilling speed is 1,400 rpm for S500 plates and 1,700 rpm for S355 plates. Drilling performance was improved with thinner plates

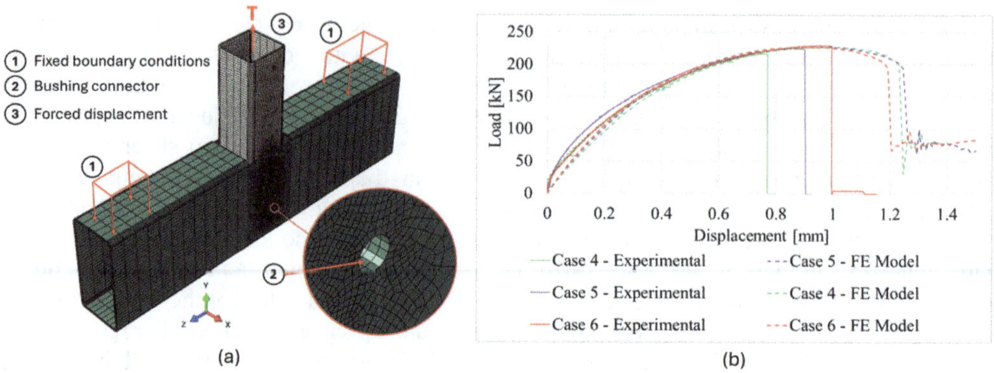

Figure 14: T-joint. (a) Numerical model; and (b) FE and experimental test force–displacement comparison.

Table 3: Ultimate force comparison between experiments and numerical models.

Cases	Experimental F_U (KN)	Numerical F_U (KN)	% Difference
04 Tension	217.02	227.22	4.5
05 Tension	225.09	227.37	1.0
06 Tension	229.82	227.23	1.2

and pre-made pilot holes. The Pias 5.5 × 32/16 screw performed best due to the length of drill bit. S500 plates required higher torque for successful drilling. With the increasing diameter of pilot holes from 3.5 mm to 4.5 mm, the drilling performance including drilling time and success rate has been improved. However, the effects of increasing pilot hole diameters on the mechanical behaviour of the joints need to be studied further.

The lap-shear tests showed that a gap distance from zero to 2 mm significantly reduces the shear strength, with a 30% drop in S355MC steel and 20% in S500MC. Ductility was slightly higher with a 2 mm gap than with zero gap. Plate arrangement also influenced the strength – 8 + 4 mm plates in S355 saw a 7% reduction, while in S500, shear strength remained unchanged. The primary failure mode was shear fracture of the screw, with slight tilting observed before failure. The plates remained less deformed, indicating that the screw bore the larger shear stresses before fracturing along the shear plane and the bearing stresses around the holes of the plates are relative smaller.

For T-joint tests, the screws failed in pure shear, with minor bending in some cases. Minor hole deformations occurred in some cases due to bearing stresses, varying in severity due to misalignment from manufacturing tolerances and drilling issues. Additionally, the T-joint tests showed that increasing screw spacing improves shear strength and ductility. Case 6 (25 mm spacing) had 6% higher shear strength and 30% greater ductility than Case 4 (15 mm spacing). Shear strength of average single screws in T-joint was lower than in lap-shear tests, likely due to joint rigidity preventing screw rotation and the gap between brace and chord walls in one side. Hole misalignments caused uneven bearing resistance but did not significantly impact the performance.

The numerical part of this study included three steps: connectors validation, lap shear tests validation and T-joint tests validation. The FE model composed of only a bushing connector accurately captured the behaviour of the experimental lap shear tests. However, stress/strain field around screw holes in the plates might differ from the reality when the bushing connectors replaced screws in complete lap shear and T-joint models. This indicates that, although using connectors in place of screw geometry provides a reasonable approximation of overall load-carrying behaviour, this simplification might not fully represent the stress/strain behaviour of steel plate around the screw holes. Therefore, further investigation is necessary to assess the suitability of employing this simplified approach, particularly for parametric studies.

ACKNOWLEDGEMENT

RFCS funding for DreamFAB project (grant number 101111547) is greatly acknowledged by the authors.

REFERENCES

[1] Zaharia, R. & Dubina, D., Stiffness of joints in bolted connected cold-formed steel trusses. *Journal of Constructional Steel Research*, **62**, pp. 240–249, 2015.
[2] Kaitila, O., Study on the behavior of a new light-weight steel roof truss. *Rakenteiden Mekaniika*, **31**(3–4), pp. 4–15, 1998.
[3] Pedreschi, R.F. & Sinha, B.P., The potential of press-joining in cold-formed steel structures. *Construction and Building Materials*, **10**(4), pp. 243–250, 1996.
[4] Szlendak, J.K. & Szpyrka, A., Resistance of tension brace in plug and play N shape RHS truss connection. *Journal of Civil Engineering, Environment and Architecture*, **65**(2/18), pp. 177–186, 2018.
[5] Szlendak J.K. & Oponowicz P.L., Experimental tests and numerical models of one and double side non-welded T RHS truss joints. *11th International Conference on Modern Building Materials, Structures and Technique*, Vilnius, 16–17 May, 2013.
[6] Szlendak, J.K., Resistance of laser made T RHS joints under compression load. *SDSS' Rio 2010 Stability and Ductility of Steel Structures*, Rio de Janeiro, Brazil, 8–10 Sep., 2010.
[7] Iriarte, C.G., Retaining system for joints. Patent number US20200011358. World Intellectual Property Organization (WIPO), 2022. https://patentscope.wipo.int.
[8] Yuce, C., Karpat, F., Yavuz, N. & Sendeniz, G., Designing for sustainability and reliability in an automotive seat structure. *Sustainability*, **6**(7), pp. 4608–4631, 2014.

SECTION 4
NATURAL FIBRE COMPOSITES AND BIOMECHANICS

EFFECT OF ALKALINE TREATMENT UNDER TENSILE STRESS ON MECHANICAL PROPERTIES OF NATURAL FIBRE REINFORCED PLA COMPOSITES

SEIJI MITSUBAYASHI & KENICHI TAKEMURA
Department of Mechanical Engineering, Kanagawa University, Japan

ABSTRACT
Recently, the development of green composites (GCs), which are 100% biomass-derived composites made of biodegradable resin reinforced with natural fibres, has attracted much attention. To improve the mechanical properties of GCs, it is necessary to enhance the adhesion between the fibre/resin interface, and alkaline treatment and cyclic tensile loading of natural fibres are considered to be effective methods. However, the optimal conditions of alkaline treatment and tensile loading for maximizing the mechanical properties of GCs have not been clarified. Therefore, in this study, to investigate the effect of alkaline treatment under tensile stress on the mechanical properties of continuous natural fibre reinforced PLA composites, continuous natural fibre reinforced PLA composites were prepared by simultaneously applying alkaline treatment and tensile loading to the continuous natural fibres and their mechanical properties were evaluated. The results showed that the mechanical properties were further improved by applying tensile load during alkaline treatment. The tensile strength of the composites approached a certain value when the tensile load during alkaline treatment exceeded a certain percentage. In conclusion, the increase in tensile strength with increasing tensile load during alkaline treatment was attributed to the decrease in yarn twist angle, suggesting that it is not necessary to apply a load exceeding a certain percentage of the breaking load during alkaline treatment.
Keywords: green composites, natural fibre, alkaline treatment, tensile stress.

1 INTRODUCTION

Recently, the development of green composites (GCs), which are 100% biomass-derived, has been attracting significant attention [1]–[3]. GC is a composite that uses biodegradable resin as the matrix and natural fibres as reinforcement. However, compared to carbon fibre reinforced plastics (CFRPs) and grass fibre reinforced plastics (GFRPs), GC has the disadvantage of weak adhesion at the fibre/resin interface [4], [5]. Therefore, improving interfacial adhesion is a crucial challenge.

Fibre surface treatment is one of the effective methods to enhance interfacial adhesion. Surface treatment techniques can be broadly classified into physical and chemical methods. A physical approach involves adding nanofillers such as carbon nanotubes (CNTs) [6] and cellulose nanofibres (CNFs) [7] to fibres or resins. However, due to their microscopic size, achieving uniform dispersion of nanofillers is challenging [8]. Non-uniform dispersion can cause localized fractures and deformations [9]. Additionally, nanofillers are expensive, which may lead to increased material costs [10]. On the other hand, chemical treatments [11] include alkaline treatment [12] and silane coupling treatment [13]. Silane coupling treatment utilizes silane compounds to form strong chemical bonds between materials, thereby improving interfacial adhesion. While this method significantly enhances adhesion, its high cost and the need for specialized equipment limit its applicability. In contrast, alkaline treatment is a method that removes lignin and hemicellulose from the fibre surface using alkaline chemicals such as sodium hydroxide. This approach has the advantage of being more cost-effective than physical treatments and silane coupling treatment, as the chemicals used are relatively

WIT Transactions on The Built Environment, Vol 216, © 2025 WIT Press
www.witpress.com, ISSN 1743-3509 (on-line)
doi:10.2495/HPSM250061

inexpensive. Therefore, alkaline treatment is particularly effective when improving interfacial adhesion while keeping costs low.

Numerous studies have been conducted to enhance the mechanical properties of GCs by applying alkaline treatment to natural fibres [14]–[16]. Matsuo et al. have demonstrated that applying cyclic tensile loads to natural fibres during alkaline treatment improves the mechanical properties of green composites (GC) [17]. Additionally, it is well known that longer fibre length generally enhances the mechanical properties of fibre reinforced plastics (FRP). Therefore, to improve the mechanical properties of GC, it is crucial to perform continuous alkali treatment on natural fibres while applying tensile loads.

Typically, natural fibres are spun into single or plied yarns before being distributed in the market [18]. However, alkali treatment of these yarns is often performed in a discontinuous manner, resulting in low productivity. To address this issue, the authors utilized a system that enables alkaline treatment while continuously applying tensile load to natural fibres, aiming to improve both the mechanical properties and production efficiency of natural fibres and green composites. Nevertheless, the effect of tensile stress variations during alkaline treatment on the mechanical properties of natural fibre-reinforced PLA composites remains insufficiently understood.

This study investigates the mechanical properties of natural fibres and green composites subjected to continuous alkaline treatment under different tensile stresses. Tensile testing was conducted to compare their mechanical performance and analyse the key factors influencing these variations.

2 EXPERIMENTAL METHOD

2.1 Materials

In this study, 20-count twisted ramie yarn (Tosco Co., Ltd., Tokyo, Japan) was used as reinforcement. It was a single-ply and right-handed twill (S-twist) [19]. In this context, the term ‘count’ refers to the metric count (Nm), which can be converted to tex, the mass per kilometre of twisted yarn, using eqn (1):

$$\mathrm{tex} = \frac{1000}{\mathrm{Nm}}, \tag{1}$$

where tex and Nm are the linear density in tex and the metric count, respectively. Therefore, the count of twisted yarns used in this study was equivalent to 50 tex. Moreover, the diameter of ramie yarn, which was observed by zoom-type stereo microscope (SZX7, Olympus Corporation, Japan), was approximately 0.3 mm.

PLA in pellet form (Terramac TE-2000, Unitika Ltd., Osaka, Japan) was used for matrix. It had a specific gravity of 1.25, a melting temperature of 170°C and a melt flow rate of 10 g/10 min (190°C, 21.18 N) [20].

In addition, sodium hydroxide (NaOH) and acetic acid (both purchased from Fujifilm Wako Pure Chemical Corporation, Osaka, Japan) were used for alkaline treatment and subsequent neutralization.

2.2 Fabrication process of specimens

2.2.1 Fabrication process of ramie yarn reinforced PLA composites

Fig. 1 shows the apparatus used for producing ramie yarn reinforced PLA filament. Alkaline treatment and compositing with PLA were conducted using this setup while feeding twisted yarns at a constant speed. The yarns were unwound for transport (part no. 1 in the Fig. 1), and rollers (part no. 2) were rotated to nip the twisted yarn. These rollers were powered by a motor, and the speed of the yarn conveyance was regulated by adjusting the rotation speed of the rollers. Tension was applied to the yarns by creating a difference in the rotational speeds of the front and rear rollers, and the tension on the yarns was measured (part no. 3). Containers for both alkaline solution and solution for neutralization (part no. 4) were set up, and alkaline treatment and neutralization were performed during transporting. In this study, to prevent hydrolysis of PLA resin, the alkaline-treated yarns were rinsed in water and thoroughly dried in a vacuum dryer at 130°C for 4 hours. The dried yarns were then conveyed into the crosshead die (part no. 7) and composited with PLA supplied from a single-shaft extruder (part no. 8). After composition, the ramie twisted yarn reinforced PLA composite was air-cooled and finally wound using a winding device (part no. 9).

① Device for unwinding twisted yarns or filament
② Rollers to nip yarn
③ Device for measuring tension applied to yarn
④ Containers for both alkaline solution and solution for neutralization
⑤ Nip twister
⑥ Far-infrared heater
⑦ Crosshead die
⑧ Single-shaft extruder
⑨ Device for winding twisted yarns or filament

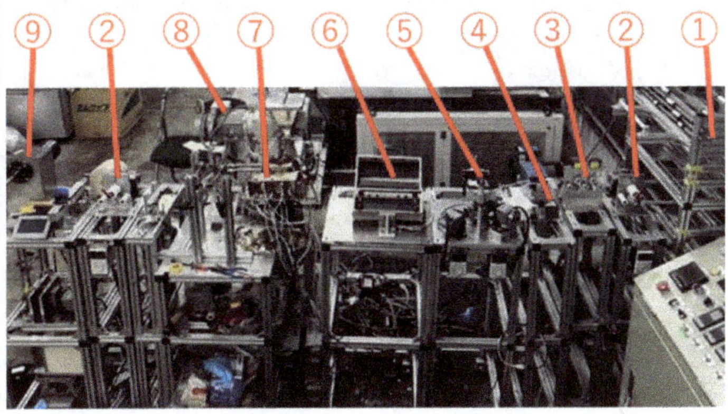

Figure 1: Equipment for the alkali treatment of ramie yarn and fabrication of ramie yarn reinforced composites.

2.2.2 Process of alkaline treatment to ramie yarn

In this study, aqueous sodium hydroxide and acetic acid solutions were used for alkaline and neutralization treatments, respectively. Fig. 2 shows the details of the container for alkaline treatment and neutralization treatment. These containers were placed at position no. 4 in Fig. 1. Inside these containers, twisted yarns were hooked to grooved bearings to reduce the frictional force acting on the twisted yarns. In addition, the length of the path that the twisted yarns travelled through the containers and the treatment time were ensured for the treatment. Here, the path length of twisted yarn passing through each container was 600 mm. When the transport speed of the twisted yarns is 18.3 mm/s, the time required for treatment is equivalent to 33 seconds. The containers were used so that the time required for alkaline treatment and neutralization treatment were equal. In this study, tension was applied to the twisted yarns during alkaline treatment at a concentration of 10 wt.% of sodium hydroxide solution. Here, the tension applied was set at 0%, 20%, 40%, 60% and 80% of the breaking load of the twisted yarns.

Figure 2: Containers for alkali treatment and neutralization.

2.2.3 Fabrication process of ramie yarn reinforced PLA composites

After alkaline treatment of the twisted yarns, ramie/PLA composites (filaments) were fabricated using the filament fabrication equipment shown in Fig. 1. The filaments were prepared by adjusting the filament diameter to 1.6 mm so that the volume content of twisted ramie yarn in the filaments was approximately 3%.

2.3 Evaluation method of tensile properties

2.3.1 Tensile test of ramie yarn

In this study, specimens were prepared by gluing alkaline-treated ramie yarns under various conditions to a U-shaped backing paper with a distance of 10 mm between the points and then cutting the backing paper with scissors. The test specimens were mounted on a universal testing machine (Tensilon, RTC-1250A, A&D Company, Limited, Tokyo, Japan) and static tensile tests were performed. The number of specimens was set to 10 or more, and the tensile speed of the specimens was set to 1.0 mm/min.

When calculating the cross-sectional area, the shape of the twisted yarn was considered to be elliptical because of the flattened shape of the fibre. When measuring dimensions, the diameters of the major and minor axes were measured by image analysis. Eqn (2) was used to calculate the cross-sectional area.

$$A = \frac{\pi ab}{4}, \tag{2}$$

where A, a, and b are the cross-sectional area, diameter of the major axis, and diameter of the minor axis of the twisted yarn, respectively.

2.3.2 Tensile test of ramie yarn reinforced PLA composites

Static tensile tests were performed on ramie/PLA composites with filament shape. Here, the length of the specimen was 150 mm, the distance between the points was 50 mm, the number of tests was 5, and the tensile speed of the specimen was 1.0 mm/min. A micrometre was used to measure the diameter.

2.4 Observation method of ramie yarn reinforced PLA composites

To investigate the factors that cause changes in the tensile strength of twisted yarns treated with alkaline under tensile stress and filaments formed using these yarns, the condition of the filaments was observed using zoom-type stereo microscope (SZX7, Olympus Corporation, Japan). The observation results were photographed using a C-mount camera for microscopy, and the twist angle of the ramie yarn inside the filament was measured. Image processing software (ImageJ) was used to measure the twist angle. Here, the number of twist angle measurements was set to 5 or more.

3 RESULTS AND DISCUSSION

3.1 Effect of alkaline treatment under tensile stress on tensile properties of ramie yarn

Fig. 3 shows the tensile strength and Young's modulus of alkali-treated ramie yarn under tensile stress. As a result of the tensile load, an increase in tensile strength and a decrease in Young's modulus were observed. Additionally, the tensile strength and Young's modulus remained largely consistent regardless of the magnitude of the tensile load applied during alkaline treatment. This suggests that the mechanical properties are independent of the tensile load during alkaline treatment.

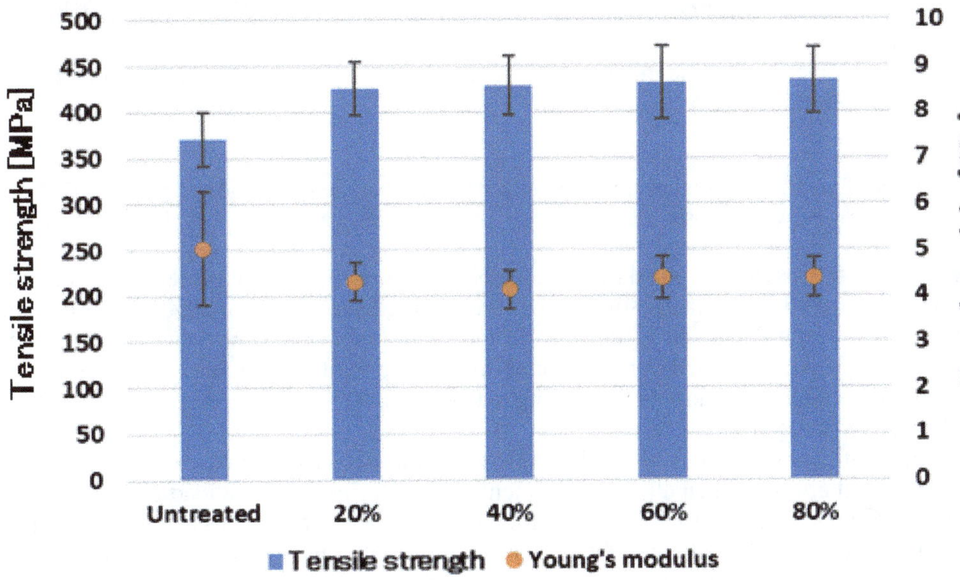

Figure 3: Effect of alkaline treatment under tensile stress on tensile properties of ramie yarn.

3.2 Effect of alkaline treatment under tensile stress on tensile properties of ramie yarn reinforced PLA composites

Fig. 4 shows the relationship between tensile stress and filament tensile strength during alkaline treatment. The maximum tensile strength was obtained when alkaline treatment was performed while applying a load of 40% of the breaking load. This result indicates that there is no advantage to alkaline treatment while applying a tensile load of 40% or greater.

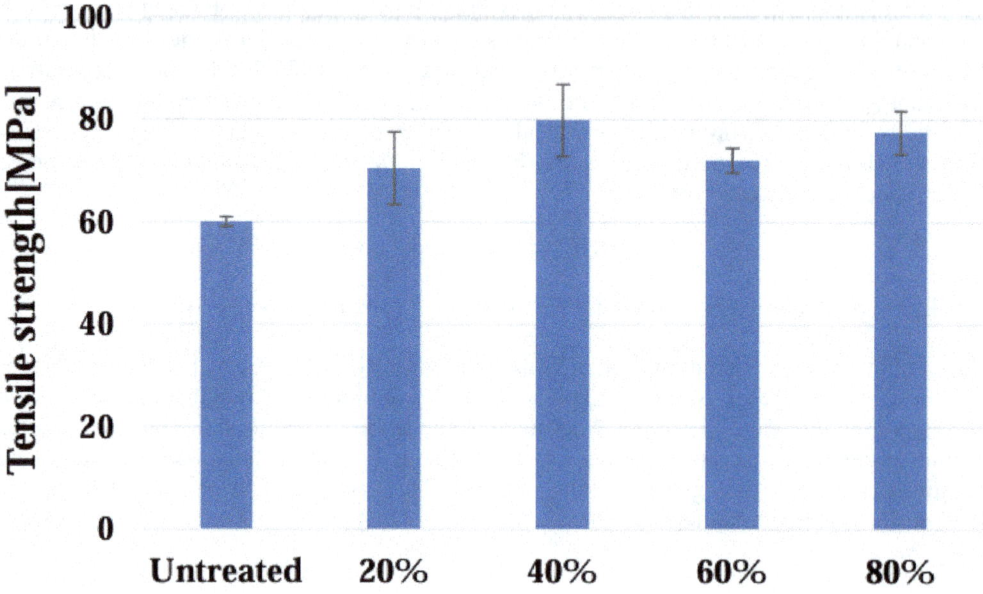

Figure 4: Effect of alkaline treatment under tensile stress on tensile strength of ramie yarn reinforced PLA composites.

3.3 Effect of alkaline treatment under tensile stress on tensile properties of ramie yarn reinforced PLA composites

Fig. 5 shows the results of optical microscope observations of filaments obtained by alkaline treatment with a 10 wt.% NaOH solution while applying tensile loads of 20%, 40%, 60% and 80% relative to the tensile strength of untreated ramie yarn. Fig. 6 presents the calculated twist angles of ramie twisted yarn subjected to alkaline treatment under a 10 wt.% NaOH solution while applying loads corresponding to 0%, 20%, 40%, 60% and 80% of the breaking load of the twisted yarn.

In all treatment conditions, the twist angle tended to decrease compared to the untreated samples. Furthermore, when alkaline treatment was performed under a load exceeding 40% of the breaking load, the twist angles were found to be approximately constant. This reduction of twist angle is considered a key factor contributing to the improvement in the strength of both twisted yarn and filaments.

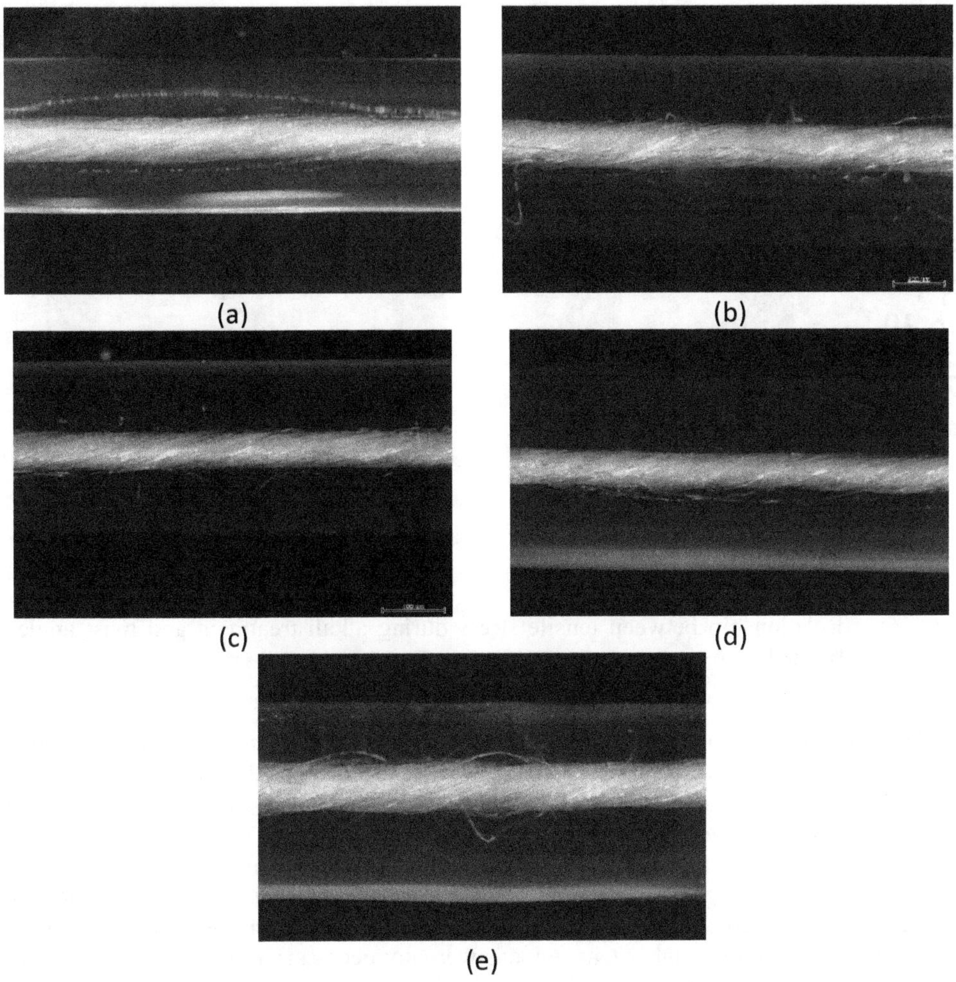

Figure 5: Filaments prepared by alkaline treatment under tensile load. (a) 0% of breaking load; (b) 20% of breaking load; (c) 40% of breaking load; (d) 60% of breaking load; and (e) 80% of breaking load.

4 CONCLUSION

The purpose of this study was to clarify the effects of changes in alkaline treatment concentration, treatment time, and tensile load on the mechanical properties of ramie/PLA composites. As a result, the following findings were obtained.

1. Measurements of the tensile strength of ramie twisted yarn showed that the tensile strength increased and Young's modulus decreased due to the applied tensile load. Additionally, the tensile strength and Young's modulus remained largely consistent regardless of the magnitude of the tensile load applied during alkaline treatment. This suggests that the mechanical properties are independent of the tensile load during alkaline treatment.

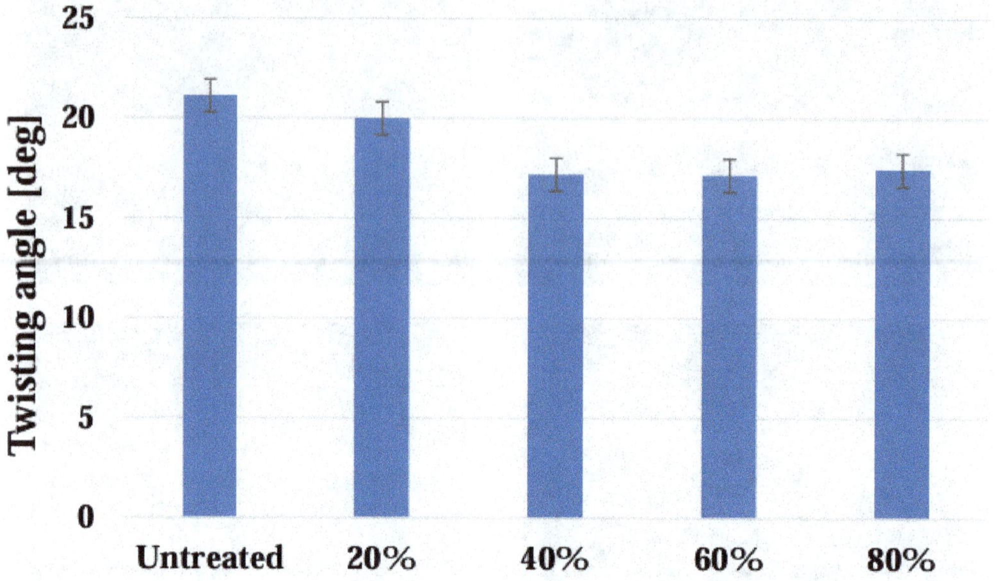

Figure 6: Relationship between tensile stress during alkali treatment and twist angle of twisted yarns.

2. The maximum tensile strength of alkali-treated ramie/PLA composites was obtained when alkaline treatment was performed while applying a load of 40% of the breaking load. This result indicates that there is no advantage to alkaline treatment while applying a tensile load of 40% or greater.
3. Tensile loading has a significant effect on the tensile strength of ramie/PLA composites. The continuous processing equipment made it possible to reduce the twist angle of the yarn. The variation of the twist angle became small when the tensile load exceeded 40%, and it was found that higher tensile loads were not necessary to increase the strength of the yarn.

ACKNOWLEDGEMENTS

We gratefully thank Mr. Matsuyama and Mr. Nagano at Kanagawa University for their support in the data acquisition.

REFERENCES

[1] Abdul Khalil, H.P.S., Bhat, A.H. & Ireana Yusra, A.F., Green composites from sustainable cellulose nanofibrils: A review. *Carbohydrate Polymers*, **87**(2), pp. 963-979, 2012.

[2] Dejene, B.K. & Geletaw, T.M., Development of fully green composites utilizing thermoplastic starch and cellulosic fibers from agro-waste: A critical review. *Polymer-Plastics Technology and Materials*, **63**(5), pp. 540–569, 2024.

[3] Somasundaram, R., Somasundaram, M., Ananchaperumal, P., Suyambulingam, I. & Siengchin, S., Enhancing properties and sustainability: Surface modified cymbopogon flexuosus stem fiber for green composite materials. *Polymer Composites*, **45**(3), pp. 2809–2824, 2024.

[4] Singh, A.P., Zafar, S., Suman, S. & Pathak, H., Mechanical and electromagnetic interference shielding properties of natural fiber reinforced polymer composite with carbon nanotubes addition. *Polymer Composites*, **45**(5), pp. 4487–4501, 2024.
[5] Teeraphantuvat, T., Jatuwong, K., Jinanukul, P., Thamjaree, W., Lumyong, S. & Aiduang, W., Improving the physical and mechanical properties of mycelium-based green composites using paper waste. *Polymers*, **16**(2), p. 262, 2024.
[6] Ali, Z., Yaqoob, S., Yu, J. & D'Amore, A., Critical review on the characterization, preparation, and enhanced mechanical, thermal, and electrical properties of carbon nanotubes and their hybrid filler polymer composites for various applications. *Composites Part C: Open Access*, **13**, 100434, 2024.
[7] Du, K., Zhang, D., Zhang, S. & Tam, K.C., Advanced functionalized materials based on layer-by-layer assembled natural cellulose nanofiber for electrodes: A review. *Small*, **20**, 2304739, 2024.
[8] Mirzapour, M., Cousin, P., Robert, M. & Benmokrane, B., Dispersion characteristics, the mechanical, thermal stability, and durability properties of epoxy nanocomposites reinforced with carbon nanotubes, graphene, or graphene oxide. *Polymers*, **16**(13), p. 1836, 2024.
[9] Wan, W., Shen, R., Tang, J., Xu, Y. & Zou, X., Carbon fiber reinforced composites with self-sensing and self-healing capabilities enabled by CNT-modified nanofibers. *Polymer Composites*, **45**(8), pp. 7301–7315, 2024.
[10] Kumar, V. & Tang, X., New horizons in nanofiller-based polymer composites II. *Polymers*, **15**(21), p. 4259, 2023.
[11] Mundhe, A. & Kandasubramanian, B., Advancements in natural fiber composites: Innovative chemical surface treatments, characterization techniques, environmental sustainability, and wide-ranging applications. *Hybrid Advances*, **7**, 100282, 2024.
[12] Sari, N.H., Sari, S., Sutaryono, Y.A. & Khan, M.A., A review: Fracture structure of natural fiber after treatment with various alkali chemicals. *Journal of Fibers and Polymer Composites*, **3**(2), pp. 158–180, 2024.
[13] Islam, T., Chaion, M.H., Jalil, M.A., Rafi, A.S., Mushtari, F., Dhar, A.K. & Hossain, S., Advancements and challenges in natural fiber-reinforced hybrid composites: A comprehensive review. *SPE Polymers*, **5**(4), pp. 481–506, 2024.
[14] Reddy, K.O., Reddy, K.R.N., Zhang, J., Zhang, J. & Rajulu, A.V., Effect of alkali treatment on the properties of century fiber. *Journal of Natural Fibers*, **10**(3), pp. 282–296, 2013.
[15] Pokhriyal, M., Rakesh, P.K., Rangappa, S.M. & Siengchin, S., Effect of alkali treatment on novel natural fiber extracted from Himalayacalamus falconeri culms for polymer composite applications. *Biomass Conversion and Biorefinery*, **14**(16), pp. 18481–18497, 2024.
[16] Dejene, B.K., False banana (*Enset Ventricosum*) fibers: An emerging natural fiber with distinct properties, promising potentials, challenges and future prospects: A critical review. *Journal of Natural Fibers*, **21**(1), 2311303, 2024.
[17] Matsuo, T., Gomes, A., Goda, K. & Ohgi, J., Improvement of mechanical properties of curaua fiber green composites. *Key Engineering Materials*, **334**, pp. 369–372, 2007.
[18] Doi, T., Ito, M., Kaji, T., Nakamura, R., Goda, K. & Ohgi, J., Improvement in strength properties of ramie fibers and ramie yarn green composites through cyclic tensile loading. *Journal of the Japan Society for Composite Materials*, **35**(2), pp. 56–63, 2009.
[19] Wang, X., Fundamentals of yarn technology. *New England University: WOOL*, **482**, p. 582, 2009.

[20] Mathurosemontri, S., Auwongsuwan, P., Nagai, S. & Hamada, H., The effect of injection speed on morphology and mechanical properties of polyoxymethylene/poly (lactic acid) blends. *Energy Procedia*, **56**, pp. 57–64, 2014.

EFFECT OF VOLUME FRACTION AND INTERFACIAL AREA BETWEEN FIBRE AND RESIN ON MECHANICAL PROPERTIES OF CNF-ADDED NATURAL FIBRE REINFORCED PLA COMPOSITES

SEIJI MITSUBAYASHI & KENICHI TAKEMURA
Department of Mechanical Engineering, Kanagawa University, Japan

ABSTRACT

In recent years, green composites (GCs), which are biodegradable resins reinforced with natural fibres and have excellent specific strength, have attracted attention from the viewpoint of environmental impact. To improve the mechanical properties of GCs, it is necessary to enhance the interfacial adhesion between fibre and resin, and recent attempts have been focused on improving the mechanical properties of GCs by adding cellulose nanofibres (CNF) to the interface between fibre and resin. However, the effects of volume fraction and interfacial area between fibre and resin on the mechanical properties of CNF-added natural fibre reinforced composites have not been clarified. Therefore, to explain these effects, the tensile strength of CNF-added natural fibre reinforced composites was evaluated by varying the content of twisted natural fibres in the composites and the interfacial area between the fibres and the resin. The following conclusions can be obtained from this study: (1) The tensile strength of natural fibre-reinforced PLA composites decreases as the interfacial area between the fibres and the resin increases, even when the fibre volume is the same; and (2) Although the tensile strength can be improved by increasing the amount of CNF as the interfacial area increases, the strength cannot be improved as much as when the interfacial area is small because the fibres tend to aggregate.

Keywords: green composites, natural fibre, volume fraction, interfacial area, cellulose nanofibre, tensile strength.

1 INTRODUCTION

In recent years, green composites (GCs) have been developed to reduce environmental impact. GCs are composites that use biodegradable resin as the matrix and natural fibres as reinforcement. They have attracted attention from a sustainability perspective [1]–[3]. However, GCs have inferior mechanical properties compared to carbon fibre reinforced plastics (CFRP) and glass fibre reinforced plastics (GFRP) [4], [5].

The main reasons for the lower mechanical properties of GCs are (1) low adhesion at the fibre/resin interface, (2) heterogeneous fibre dispersion, and (3) degradation due to the hygroscopic properties of natural fibres [6], [7]. In particular, low interfacial adhesion causes fibre pullout and interface debonding under tensile loading, resulting in a decrease in mechanical strength [8]. Therefore, it is necessary to reduce the area of the fibre/resin interface when designing GC.

On the other hand, in practice, there are cases where the interface area needs to be increased. For example, when the fibre content is increased to increase the percentage of reinforcement and the overall strength [9], an increase in the interfacial area is inevitable. The interfacial area also needs to be increased when there is a limit to the fibre diameter when moulding GC. Therefore, if the area of the fibre/resin interface needs to be increased, appropriate interfacial strengthening techniques are essential.

Chemical surface modification techniques such as alkali treatment [10] and silane coupling treatment [11] have been proposed as methods to improve interfacial adhesion. Silane coupling treatment is a technique that improves interfacial adhesion by forming strong

WIT Transactions on The Built Environment, Vol 216, © 2025 WIT Press
www.witpress.com, ISSN 1743-3509 (on-line)
doi:10.2495/HPSM250071

chemical bonds between materials using silane compounds. Alkali treatment is a method of removing lignin and hemicellulose from the fibre surface using alkaline chemicals such as sodium hydroxide. However, these chemical treatments are costly and have a large environmental impact [12]. Therefore, there is a need for an interface strengthening method with a lower environmental burden.

In recent years, the reinforcement of interfaces using cellulose nanofibres (CNF) has gained attention. CNF is a naturally derived material with a nanoscale high specific surface area and low environmental impact [13]. It is expected to function as a nanoscale reinforcing agent, enhancing the mechanical strength of the interface [14], [15]. Therefore, to mitigate the deterioration of mechanical properties caused by increased interfacial area and to ensure stable strength characteristics, CNF-based interfacial reinforcement is considered effective. However, the effect of CNF addition on the mechanical properties of natural fibre-reinforced PLA composites is assumed to change depending on fibre volume fraction and fibre/matrix interfacial area. The influence of these parameters on the mechanical properties of CNF-reinforced natural fibre/PLA composites has not yet been fully clarified.

Therefore, in this study, multiple specimens were prepared with varying fibre volume fractions and fibre/matrix interfacial areas to investigate their effects on the mechanical properties of natural fibre-reinforced PLA composites. Furthermore, CNF was added at different concentrations to specimens with modified interfacial areas, and the changes of interfacial adhesion were compared.

2 EXPERIMENTAL METHOD

2.1 Materials

In this study, four types of twisted ramie yarns were used as reinforcing materials. Among them, the 7, 20, and 60 count twisted yarns were all obtained from Tosco Co., Ltd. and Tokyo, Japan, while only the 40 count was obtained from Bookbinding Studio Marumizu-Gumi, Tokyo, Japan. In this context, the term 'count' refers to the metric count (Nm), which can be converted to tex, the mass per kilometre of twisted yarn, using eqn (1):

$$\mathrm{tex} = \frac{1000}{\mathrm{Nm}}, \tag{1}$$

where tex and Nm are the linear density in tex and the metric count, respectively.

Polylactic acid (PLA) (TERRAMAC TE-2000, UNITIKA Ltd., Osaka, Japan) was used as the base material, which had a specific gravity of 1.25, a melting temperature of 170°C, and a melt flow rate of 10 g/10 min (190°C, 21.18 N) [16].

Additionally, CNF (BiNFi-s, WFo-10002, Sugino Machine Co. Ltd., Japan) was used as an additive. The CNF used in this study is commercially available in slurry form, dissolved at a concentration of 2 wt. To incorporate CNF into the ramie twisted yarns, the CNF slurry was first diluted with distilled water and then stirred using a homogenizer.

2.2 Fabrication process of specimens

2.2.1 Fabrication process of CNF-added ramie yarn reinforced PLA composites

Fig. 1 shows the apparatus for producing ramie twisted yarn reinforced PLA filament. Alkali treatment and compositing with PLA were performed using this apparatus while feeding twisted yarns at a constant speed. The twisted yarns were unwound for transport (part no. 1 in Fig. 1) and the rollers were rotated while nipping the twisted yarns (part no. 2). A motor was attached to each roller, and the twisted yarns were conveyed by adjusting the rotation speed of each roller. Here, the twisted yarns were conveyed while making sure that tension was applied to the twisted yarns so that they would not sag (part no. 3). When adding CNF to the conveyed twisted yarns, CNF was adhered to the ramie twisted yarns while feeding CNF slurry into a container (part no. 4) using a roller tube pump. Here, the CNF was agitated beforehand in a process homogenizer to fully agitate the CNF in the slurry. The rotational speed and operation time of the homogenizer during agitation were set to 2500 rpm and 30 minutes, respectively. In this study, to inhibit hydrolysis of PLA resin, twisted yarns with CNF were thoroughly dried in a vacuum dryer. Here, the temperature and time for drying were 130°C and 4 hours, respectively. After drying, the CNF-added ramie twisted yarns were conveyed into the crosshead die (part no. 7) and composited with PLA supplied by a single-shaft extruder (part no. 8). The composited ramie twisted yarn reinforced PLA composite was air cooled and finally wound while using a winding device (part no. 9).

① Device for unwinding twisted yarns or filament
② Rollers to nip yarn
③ Device for measuring tension applied to yarn
④ Container for adding CNF
⑤ Nip twister
⑥ Far-infrared heater
⑦ Crosshead die
⑧ Single-shaft extruder
⑨ Device for winding twisted yarns or filament

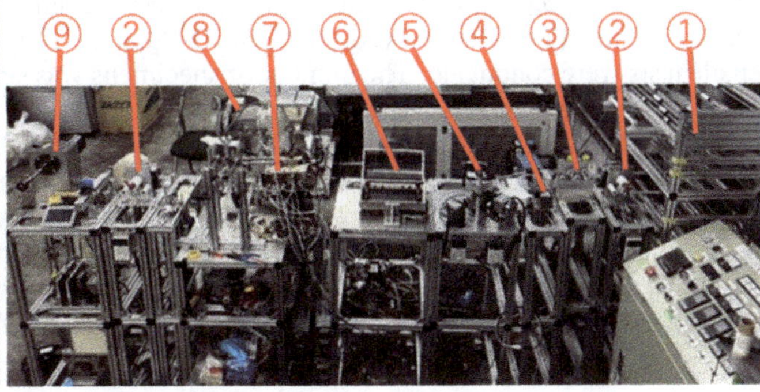

Figure 1: Equipment for the fabrication of CNF-added ramie yarn and ramie yarn reinforced composites.

2.2.2 Fabrication process of ramie yarn reinforced PLA composites

CNF-added ramie/PLA composites (filaments) were fabricated using the filament fabrication equipment shown in Fig. 1. The filament fabrication conditions are shown in Table 1 and Table 2. The filaments were fabricated while adjusting the diameter of the fabricated filaments to 1.6 mm. The filaments were also fabricated while adjusting the mass of CNF to be added in each condition in Table 2 between 0% and 1% of the mass of the slurry at 0.25% intervals. Here, when adding CNF, each twisted yarn was immersed in the slurry individually rather than immersing multiple yarns simultaneously.

Table 1: Fabrication conditions of filaments to evaluate the relationship between tensile strength and volume fraction of filaments.

No.	Count (Nm)	Number of yarns	Volume fraction (%)	CNF additive ratio (wt.%)
1	7	1	11.6	0
2	20	1	4.42	0
3	40	1	2.25	0
4	60	1	1.46	0

Table 2: Fabrication conditions of filaments for evaluating the relationship between tensile strength and fibre-resin interfacial area.

No.	Count (Nm)	Number of yarns	Volume fraction (%)	CNF additive ratio (wt.%)
1	20	1	4.42	0, 0.25, 0.50, 0.75, 1.0
2	40	2	4.50	0, 0.25, 0.50, 0.75, 1.0
3	60	3	4.38	0, 0.25, 0.50, 0.75, 1.0

2.3 Evaluation method of tensile properties

2.3.1 Tensile test of ramie yarn

In this study, the four types of ramie yarns shown in Table 1 were glued to a U-shaped backing paper with a 10 mm distance between the points, and the backing paper was then cut with scissors to prepare the test specimens. The test specimens were mounted on a universal testing machine (Tensilon, RTC-1250A, A&D Company, Limited, Tokyo, Japan) and static tensile tests were conducted. The number of specimens was set to 10 or more, and the tensile speed of the specimens was set to 1.0 mm/min.

When calculating the cross-sectional area, the shape of the twisted yarn was considered to be elliptical because of the flattened shape of the fibre. When measuring dimensions, the diameters of the major and minor axes were measured by image analysis. Eqn (2) was used to calculate the cross-sectional area.

$$A = \frac{\pi ab}{4}, \quad (2)$$

where A, a, and b are the cross-sectional area, diameter of the major axis, and diameter of the minor axis of the twisted yarn, respectively.

2.3.2 Tensile test of ramie yarn reinforced PLA composites

Static tensile tests of ramie/PLA composites in filament shape were performed under each of the conditions listed in Table 1 and Table 2. Here, the length of the specimen was 150 mm, the distance between the points was 50 mm, the number of tests was 5, and the tensile speed of the specimen was 1.0 mm/min. A micrometre was used to measure the diameter.

2.4 Observation of fracture surface of ramie yarn reinforced PLA composite

A scanning electron microscope (SEM) (SU8010, Hitachi, Limited, Tokyo, Japan) was used to observe the fracture surface of the filaments. The acceleration voltage was set to 5.0 kV and the magnification to 250x to observe the shape of the fracture surface and voids. To observe the interface between the twisted yarns and resin, the acceleration voltage was set to 5.0 kV and the magnification was 500x.

3 RESULTS AND DISCUSSION

3.1 Tensile strength of ramie yarns

Fig. 2 shows the average tensile strength of 7-count, 20-count, 40-count, and 60-count ramie yarns, each measured five times. The values ranged from 235 to 260 MPa, with an average of approximately 250 MPa. Since all three yarn types exhibited similar strength, they can be considered equivalent regardless of yarn thickness.

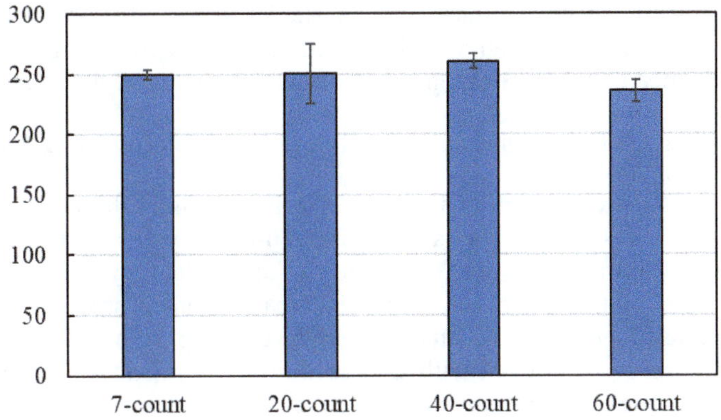

Figure 2: Tensile strength of ramie yarns.

3.2 Tensile strength of ramie yarn reinforced PLA composites

Fig. 3 shows a comparison between the tensile strength of filaments containing a single fibre for each of the 7, 20, 40, and 60 count yarns, as listed in Table 1, and the theoretical values calculated based on the rule of mixtures.

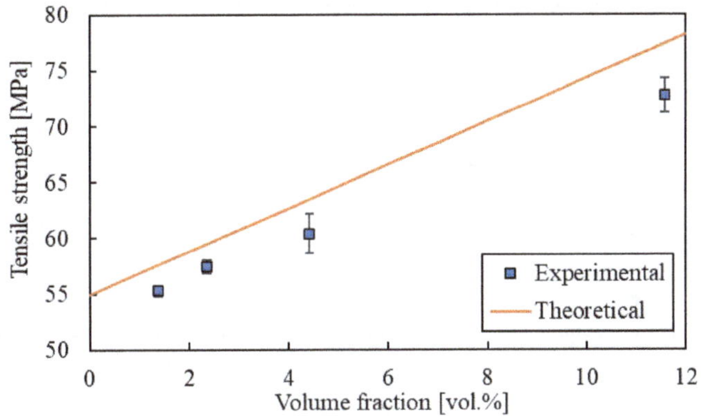

Figure 3: Relationship between tensile strength of ramie yarn reinforced PLA composites and volume fraction of yarn.

When applying the rule of mixtures, the load was assumed to act in the fibre direction. Therefore, the tensile strength of the FRP was determined as the volume-weighted average of the fibre and resin strengths, expressed as follows:

$$\sigma_c = \sigma_f V_f + \sigma_m V_m, \tag{3}$$

where σ represents the tensile strength, V denotes the volume fraction, and the subscripts c, m, and f refer to the composite, matrix, and fibre (yarn), respectively.

As shown in Fig. 3, the experimental values tended to be lower than the theoretical values calculated using the rule of mixtures. The failure of FRP can be attributed to multiple factors, including fibre fracture, matrix failure, and interfacial debonding, which indicate that the actual strength does not always align with idealized theoretical predictions. Nevertheless, the results suggest the need for improvements in the tensile strength of the filament.

3.3 Effect of adding CNF on tensile strength of ramie yarn reinforced PLA filament

Fig. 4 presents the tensile strength of filaments fabricated under three different conditions: one 20-count yarn, two 40-count yarns, and three 60-count yarns. In these specimens, the volume fraction of ramie yarn was adjusted to approximately 4.4%, allowing for a comparison of the relationship between interfacial area and tensile strength. Additionally, the figure also illustrates the changes of tensile strength resulting from CNF addition. As figure 4 shows, it was observed that even with CNF addition, the maximum tensile strength was obtained under the one 20-count yarn condition. A comparison of the increase in tensile strength between 0 wt.% and 0.25 wt.% CNF addition showed that the highest increase was observed in the three count-60 yarns condition, followed by the two count-40 yarns and one count-20 yarn conditions. In particular, the condition with three 60-count yarns showed the greatest improvement in tensile strength. This is likely because it had the largest interfacial area, making it more responsive to the effects of CNF. However, at a CNF slurry concentration of 0.50 wt.%, a decrease in tensile strength was observed only in the condition with three 60-count yarns. This may be attributed to excessive CNF addition, leading to aggregation and a loss of interfacial uniformity.

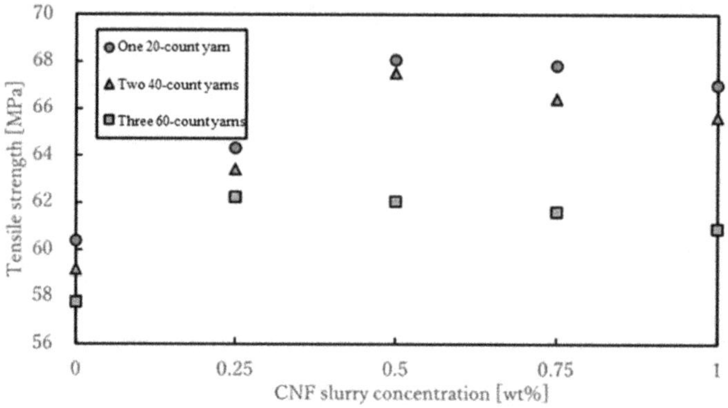

Figure 4: Relationship between tensile strength and CNF slurry concentration.

Furthermore, theoretical values were calculated using eqn (2) based on the rule of mixtures presented in Section 3.2, yielding a theoretical tensile strength of 63.5 MPa. A comparison between this theoretical value and the maximum tensile strength of the filaments under each condition was conducted to analyse the effect of CNF addition. As a result, under the conditions of using one 20-count yarn and two 40-count yarns, the measured tensile strength exceeded the theoretical value. In contrast, the measured tensile strength was lower than the theoretical value when using three 60-count yarns.

Scanning electron microscope (SEM) was used to observe the fracture surfaces of each filament. The observation results are shown in Fig. 5. In this figure, (a), (b), and (c) represent the fracture surfaces of the specimens obtained when 0.75 wt.% CNF slurry was added to 20-

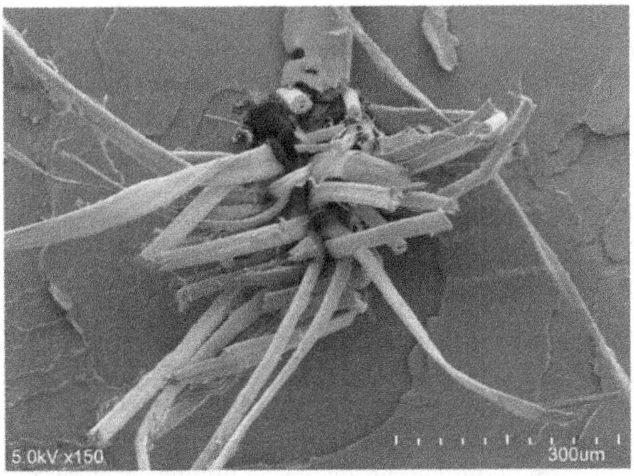

(a)

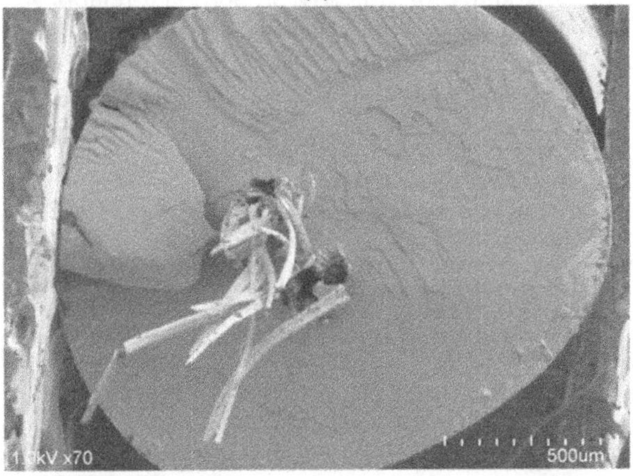

(b)

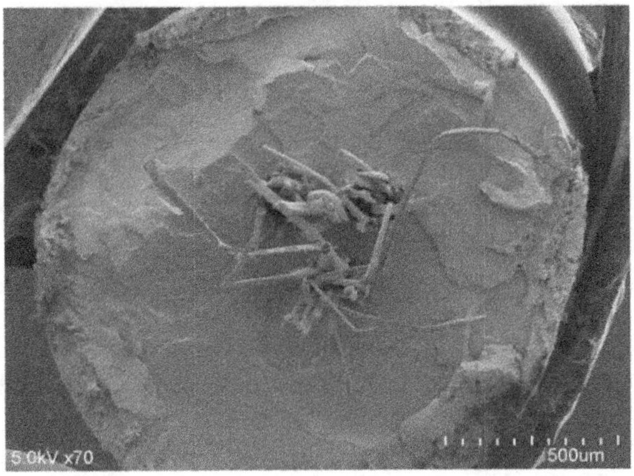

(c)

Figure 5: SEM images of ramie yarn reinforced PLA Filaments. (a) One 20-count yarn; (b) Two 40-count yarns; and (c) Three 60-count yarns.

count ramie twisted yarn, 0.75 wt.% CNF slurry was added to 40-count ramie twisted yarns, and 0.50 wt.% CNF slurry was added to 60-count ramie twisted yarns, respectively. As seen in Fig. 5(c), the weak interfacial adhesion likely caused fibre pull-out, resulting in a decrease in tensile strength. Similarly, Figs 5(a) and (b) show that weak interfacial adhesion led to fibre pull-out and a corresponding decrease in tensile strength.

3.4 Relationship between tensile strength and CNF content per unit length of CNF-added ramie yarn reinforced PLA composites

Fig. 6 shows the relationship between the tensile strength of filaments and the amount of added CNF. The filaments were produced under three conditions: one 20-count yarn, two 40-count yarns, and three 60-count yarns. The x-axis in Fig. 6 represents the CNF content per unit length of the filament. As shown in the Fig. 6, while the CNF content per strand decreases as the yarn diameter decreases, the CNF content per unit length of the filament tends to increase with a higher yarn count and an increased number of strands. Furthermore, when CNF was added to the 20-count yarn, a nearly proportional relationship between the CNF addition rate and the adhered amount was observed up to a CNF slurry concentration of 0.50 wt.%. However, beyond 0.50 wt.%, the adhesion ratio of CNF tended to decrease. Similarly, for both the 40-count and 60-count yarns, an increase in the CNF content resulted in a decrease in the adhesion ratio. By comparing these results with the tensile strength of the filaments, it was suggested that the decrease in filament strength was primarily due to CNF aggregation.

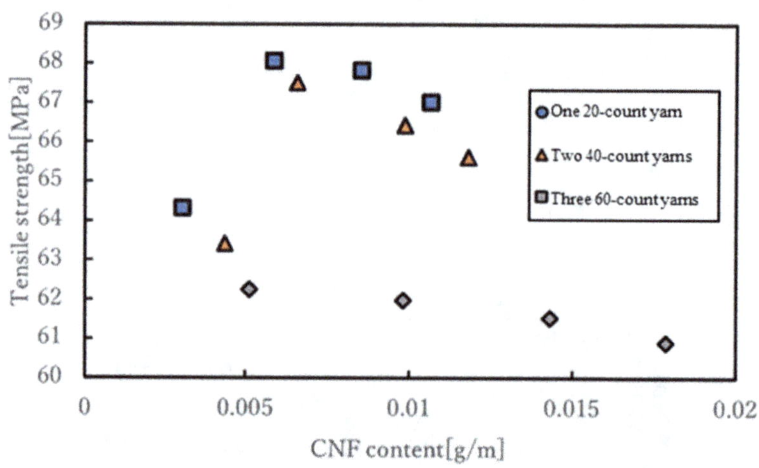

Figure 6: Relationship between tensile strength and CNF content.

When comparing the cross-sectional area per unit length of each twisted yarn, the cross-sectional areas of the 40-count and 60-count yarns are 1/2 and 1/3, respectively, of that of the 20-count yarn. Consequently, the diameters of the 40-count and 60-count yarns are$1/\sqrt{2}$ and $1/\sqrt{3}$, respectively. Thus, when comparing individual yarns, the interfacial surface area between the yarn and the resin follows the same ratio, with the 40-count and 60-count yarns having $1/\sqrt{2}$ and $1/\sqrt{3}$ of the interfacial area of the 20-count yarn. However, when considering the total interfacial area per unit length of the filament, the presence of two 40-

count yarns and three 60-count yarns results in an interfacial area that is $\sqrt{2}$ and $\sqrt{3}$ times that of a filament containing a single 20-count yarn, respectively. Although an increase in the interfacial bonding area between the yarn and the resin was expected to enhance interfacial reinforcement by CNF, in reality, the reduction in the interfacial bonding area per individual yarn led to increased CNF aggregation, which in turn contributed to a decrease in tensile strength.

4 CONCLUSION

In this study, tensile tests were conducted on composite materials to evaluate changes in tensile strength due to differences in the volume fraction of ramie yarn in the filaments and changes in tensile strength when the number of fibres was changed while maintaining the same volume fraction of ramie yarn in the filaments. The same evaluation was also conducted when CNF was added. The following are the findings obtained from these evaluations.

1. As the volume fraction of ramie twisted yarn in the filament increases, the tensile strength of the filament improves. Additionally, the tensile strength of each filament without CNF addition falls below the theoretical values calculated based on the rule of mixtures.
2. When the volume fraction of ramie twisted yarn in the filament is constant, a decrease in the yarn diameter leads to a reduction in tensile strength. This tendency is independent of CNF addition. Therefore, increasing the yarn diameter within the filament is necessary.
3. The addition of CNF enhances the tensile strength of filaments made with 20-count and 40-count twisted yarns, exceeding the theoretical values calculated based on the rule of mixtures. In contrast, for filaments produced using 60-count twisted yarn, the tensile strength does not surpass the theoretical values, even with CNF addition.
4. When the concentration of CNF slurry is 0.25 wt.% or lower, the tensile strength of the filament increases at a higher rate as the diameter of the twisted yarn decreases. This is because reducing the yarn diameter increases the interfacial area between the yarn and the resin, enhancing the tensile strength. However, when the CNF slurry concentration exceeds 0.25 wt.%, specimens with smaller yarn diameters tend to exhibit a lower rate of tensile strength improvement. This is likely due to the increased tendency of CNF aggregation.
5. The tensile strength of the filament depends not on the total interfacial area between the twisted yarn and resin but on the interfacial area per individual yarn. A decrease in the adhesion area at the yarn/resin interface for each yarn increases the likelihood of CNF aggregation, which contributes to this effect.

ACKNOWLEDGEMENTS

Some equipment was used in this study at the Research Institute for Engineering at Kanagawa University. We gratefully thank Mr. Miyase at Kanagawa University for his support in data acquisition.

REFERENCES

[1] Abdul Khalil, H.P.S., Bhat, A.H. & Ireana Yusra, A.F., Green composites from sustainable cellulose nanofibrils: A review. *Carbohydrate Polymers*, **87**(2), pp. 963–979, 2012.

[2] Dejene, B.K. & Geletaw, T.M., Development of fully green composites utilizing thermoplastic starch and cellulosic fibers from agro-waste: A critical review. *Polymer-Plastics Technology and Materials*, **63**(5), pp. 540–569, 2024.

[3] Somasundaram, R., Somasundaram, M., Ananchaperumal, P., Suyambulingam, I. & Siengchin, S., Enhancing properties and sustainability: Surface modified cymbopogon flexuosus stem fiber for green composite materials. *Polymer Composites*, **45**(3), pp. 2809–2824, 2024.

[4] Liu, J.L., Pham, V.N.H., Mencattelli, L., Chew, E., Chua, P.Y., Shen, J., Tian, K., Zhi, J., Jiang, D., Tay, T.E. & Tan, V.B.C., Improving the impact performance of natural fiber reinforced laminate through hybridization and layup design. *Composites Science and Technology*, **251**(26), 110585, 2024.

[5] Savas, S., Bakir, D. & Akcaat, Y.K., Experimental and numerical investigation of the usability of nonwoven hemp as a reinforcement material. *Case Studies in Construction Materials*, **20**, e03091, 2024.

[6] Singh, A.P., Zafar, S., Suman, S. & Pathak, H., Mechanical and electromagnetic interference shielding properties of natural fiber reinforced polymer composite with carbon nanotubes addition. *Polymer Composites*, **45**(5), pp. 4487–4501, 2024.

[7] Teeraphantuvat, T., Jatuwong, K., Jinanukul, P., Thamjaree, W., Lumyong, S. & Aiduang, W., Improving the physical and mechanical properties of mycelium-based green composites using paper waste. *Polymers*, **16**(2), p. 262, 2024.

[8] Krishnasamy, S., Kumar, M.H., Parameswaranpillai, J., Rangappa, S.M. & Siengchin, S., *Interfacial Bonding Characteristics in Natural Fiber Reinforced Polymer Composites: Fiber-Matrix Interface In Biocomposites*, Springer: Singapore, pp.1–22, 2024.

[9] You, Y.J., Kim, J.H.J., Kim, S.J. & Park, Y.H., Methods to enhance the guaranteed tensile strength of GFRP rebar to 900 MPa with general fiber volume fraction. *Construction and Building Materials*, **75**, pp. 54–62, 2015.

[10] Sari, N.H., Sari, S., Sutaryono, Y.A. & Khan, M.A., A review: Fracture structure of natural fiber after treatment with various alkali chemicals. *Journal of Fibers and Polymer Composites*, **3**(2), pp. 158–180, 2024.

[11] Islam, T., Chaion, M.H., Jalil, M.A., Rafi, A.S., Mushtari, F., Dhar, A.K. & Hossain, S., Advancements and challenges in natural fiber-reinforced hybrid composites: A comprehensive review. *SPE Polymers*, **5**(4), pp. 481–506, 2024.

[12] Mohammed, M., Mohammed, A.M., Oleiwi, J.K., Adam, T., Betar, B.O., Gopinath, S.C.B., Ihmedee, F.H. & Anton, K., Influence of fiber treatments on water absorption behavior in natural fiber-reinforced polymer composites: A review. *Al-Rafidain Journal of Engineering Sciences*, **3**(1), pp. 410–430, 2025.

[13] Du, K., Zhang, D., Zhang, S. & Tam, K.C., Advanced functionalized materials based on layer-by-layer assembled natural cellulose nanofiber for electrodes: A review. *Small*, **20**, 2304739, 2024.

[14] Sarr, M.M. & Kosaka, T., Effect of cellulose nanofibers on the fracture toughness mode II of glass fiber/epoxy composite laminates. *Heliyon*, **9**(2), e13203, 2023.

[15] Wu, D., Xing, Y., Liu, L., Dong, Q., Wang, M. & Zhang, R., Structural design of 'straw and clay' based on cellulose nanofiber/polydopamine and its interfacial stress dissipation mechanisms. *International Journal of Biological Macromolecules*, **283**(4), 138040, 2024.

[16] Mathurosemontri, S., Auwongsuwan, P., Nagai, S. & Hamada, H., The effect of injection speed on morphology and mechanical properties of polyoxymethylene/poly (lactic acid) blends. *Energy Procedia*, **56**, pp. 57–64, 2014.

MECHANICAL BEHAVIOUR OF PRINTABLE SPLINT POROUS DESIGNS USING ADDITIVE MANUFACTURING: A FINITE ELEMENT ANALYSIS

HASSAN MEHBOOB[1] & ABDELHAK OULDYEROU[2]
[1]Prince Sultan University, Saudi Arabia
[2]West Virginia University, Morgantown, USA

ABSTRACT

This study investigates the mechanical performance of splints with soft resin and acrylic biomaterials. The dental splint was created with a 3 mm height and divided into three layers. The top and bottom layers (external layers) differed by 1 and 2 mm. Body-centred cubic structures (BCC) were used to create a porous middle layer, which also varied by 1 and 2 mm. The top rigid layer was subjected to a compressive displacement in small steps to visualize the stress–strain plots for linear and non-linear regions. Results indicated that the larger the strut thickness (lower porosity) the lower stresses and reaction forces, regardless of the thickness of the middle layer. Additionally, variations in the thickness of the middle layer resulted in noticeable changes in both stresses and reaction forces. An increase in strut thickness led to higher stress levels, indicating an enhanced load-bearing capacity under the same applied deformation. In conclusion, incorporating porous layers within the splints can effectively distribute reaction forces and mechanical stresses. This design approach can offer several advantages, including enhancing patient comfort, reducing pressure on tissues such as the gingiva and peri-implant contact area.

Keywords: printable dental splints, finite element analysis, biomechanics, porous lattice structure.

1 INTRODUCTION

Restorations made of prosthetics are subject to a variety of biological and biomechanical effects. Bone loss from peri-implant infections of soft and hard tissue brought on by poor dental hygiene might be exacerbated in cases of low bone quality or increased sensitivity to infections [1]. In a similar vein, large biomechanical loads may result in excessive mechanical stress in the surrounding bone, which may lead to bone loss [2]. The distribution, direction, and magnitude of occlusal pressures acting on the implant, surrounding bone, and prosthetic reconstruction are critical factors for implant survival [3]. Bruxism patients receiving implant-supported prostheses cause parafunctional contacts and occlusal overload of the bone-implant interface [4]. However, clinical experience indicates that these individuals may develop mechanical and biological problems as a result of occlusal overload [5]. Due to their lack of a periodontal ligament and the absence of the proprioceptive feedback mechanisms of the masticatory muscles, dental implants are firmly attached to the bone. Therefore, marginal bone loss may be triggered by occlusal loading. The lifetime of an implant-supported prosthesis and the amount of bone loss surrounding an implant may also be correlated with the length and depth of implantation. This affects long-term osseointegration and is caused by variations in implant insertion depths, implant lengths, and contact areas of varied bone densities.

The art and science of employing detachable devices to produce a mechanical disadvantage for parafunctional forces and to attain neuromuscular harmony in the masticatory system is known as splint therapy [6]. Splints are advised for the therapeutic management bruxism patients in order to reduce the forces and lessen tooth wear [7]. The goal of intraoral occlusal splints is to achieve a balanced and consistent occlusal contact without irreversibly altering the dental occlusion or the mandibular rest position.

WIT Transactions on The Built Environment, Vol 216, © 2025 WIT Press
www.witpress.com, ISSN 1743-3509 (on-line)
doi:10.2495/HPSM250081

Furthermore, a well-made splint encourages a harmonious interaction between the implant, bone, teeth, ligaments, joints, disk assemblies, masticatory muscles, and tendons. The justification for their employment is that durable, soft material exhibits a shock-absorbing [8] ability that may disperse severe loads linked to parafunctional behaviours. Since occlusal devices are usually worn at night, when parafunction episodes are most likely, a good fit is necessary to keep the device from moving while the patient is asleep. Furthermore, the anterior guiding should remove contact between the occlusal device and the posterior teeth during protrusion and eccentric movement, and there should be uniform occlusal contacts between the antagonist teeth and the device at maximum intercuspation. Given that patients typically store their occlusal devices in moist storage containers while not in use, a device must be precise and resistant to environmental changes in order to produce and maintain these harmonious contacts and disocclusion, regardless of the manufacturing process. Despite the fact that clinical practitioners generally agree that the occlusal splint is the most popular treatment, there is still debate regarding the ideal vertical thickness and the thorough criteria used to determine this thickness. According to published research, splints ranging in thickness from 1 to 15 mm have been used [9]. The efficacy of splints varying in thickness on symptoms related as well. As a result, choosing the right vertical thickness can have a significant impact on how well the treatment works.

The flask–pack–press method used to fabricate traditional dentures is similar to that used to fabricate occlusal devices. Although several materials can be employed in the traditional production of occlusal devices, the procedure is costly and time-consuming. Occlusal devices can now be manufactured both additively and thanks to recent developments in computer-aided design and computer-aided manufacturing (CAD–CAM) technologies. CAD–CAM techniques are highly efficient in terms of time, cost, and reproducibility. Occlusal devices made from prepolymerized [8] and standardized disks with superior mechanical qualities to those made traditionally can be produced by subtractive manufacturing. Nonetheless, additive manufacturing (AM) offers the advantage of producing several devices at once and producing less waste than subtractive manufacturing (SM). In many different biomedical domains, this method has become widely accepted. AM approaches are highly attractive for a range of applications because to their diversified methodologies and large range of materials with different physical properties. Furthermore, the design flexibility it offers makes it possible to produce intricate structures that would be challenging to build using conventional subtractive manufacturing techniques. As a result, digital occlusal splints are frequently made using 3D printing.

Finite element analysis (FEA) techniques have been used widely to evaluate the stress concentration pattern at the bone implant interface [10], [11]. In dentistry, there is growing interest in using occlusal load models based on FEA to examine various methods and materials [12], [13]

This study aims to design and investigate the behaviour of various splints design with porous architectures employing finite element method.

2 MATERIALS AND METHODS

2.1 Finite element modelling

Dental splints section's 3D finite element models were created using SolidWorks 2023 (SolidWorks Corp., Dassault Systèmes, France). The splints' overall thickness, including their solid and porous layers, was measured at 3 mm, as shown in Fig. 1. The solid upper and lower layers, as well as the porous inner splint layers, were designed to absorb forces and

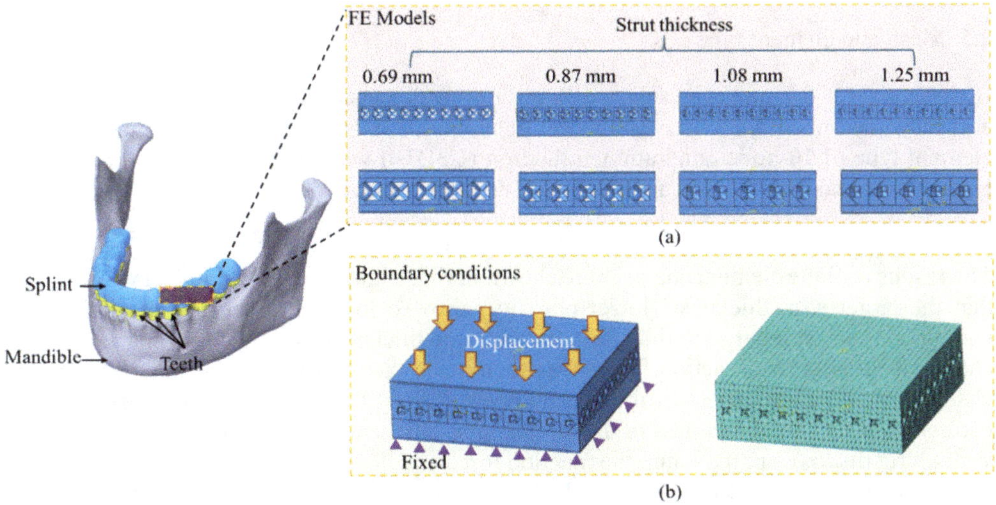

Figure 1: (a) Different splint models; and (b) mesh and boundary conditions.

reduce strains on the interfaces with the crown and between the implant and bone as well. The splints' top and bottom layers differed by 1 and 2 mm. Body-centred cubic structures (BCC) were used in the design of the middle layers, which also varied by 1 and 2 mm.

2.2 Materials assignment

Two types of biomaterials [14] were utilized to assign the splint model. Soft resin was used for the internal layer (middle layer), while acrylic was used to model the external layer (top and bottom layers). The material properties of the splints were listed in Table 1. Non-linear material properties were assigned to the analysis to accurately capture the mechanical behaviour of the splints.

Table 1: Material properties used for finite element investigation.

Material	Young's modulus (MPa)	Poisson's ratio	Yield strength (MPa)
Acrylic	2,400	0.35	110
Soft resin	1,100	0.30	45

2.3 Interactions

The top and bottom plates were modelled as rigid bodies and were connected to the splint surfaces using surface-to-surface tie constraints.

2.4 Boundary conditions

The top rigid layer was subjected to a compressive displacement of 0.5 mm and the bottom rigid layer was fixed in all directions. The compression was applied in small steps to visualize the stress–strain plots for linear and non-linear regions.

2.5 Mesh and element type

A mesh size of 0.2 mm was selected based on mesh sensitivity analysis to ensure results accuracy. Ten-node quadratic tetrahedral elements (C3D10) were used for meshing. The element type of 10-node quadratic tetrahedron (C3D10) was selected. The total number of elements ranged from 342301 to 109164 for different models.

3 RESULTS

The results of finite element analyses are shown in Fig. 2 and listed in Table 2. It was revealed that the larger strut thickness (lower porosity) showed lower stresses and reaction forces compared to the smaller strut thickness for the same thickness of the middle layer. Also, the change in stresses and reaction forces was seen when the thickness of the middle layer was changed. A higher thickness of the middle layer showed lesser stresses and reaction forces compared to a smaller thickness of the middle layer. In the case of a 1 mm thick middle layer with a strut thickness of 0.69 mm, the reaction force of 839 N was seen, however, this reaction force was increased to 1449, 2679 and 4154 N for the strut thicknesses of 0.87, 1.08 and 1.25 mm, respectively. Hence, a larger thickness of the strut can carry more load compared to a smaller thickness of the strut. Similarly, the von-Mises stresses in the case of a 1 mm thick middle layer with a strut thickness of 0.69, 0.87, 1.08 and 1.25 mm were seen as 116.9, 119.7, 121.2 and 144.2 MPa, respectively. The stresses were increased by increasing the strut thickness, which is evidence of carrying more load with the same amount of applied deformation. Similarly, in the case of a 2 mm thick middle layer, the stresses and reaction forces were 99, 74.7, 108.2, 147.9 MPa and 810, 1383.3, 2457 and 3840 N, respectively, for the strut thickness of 0.69, 0.87, 1.08 and 1.25 mm.

Table 2: Reaction forces and stresses in different splint models.

Layer thickness (mm)	Strut thickness (mm)	Stresses (MPa)	Reaction force (N)
1	0.69	116.9	839
	0.87	119.7	1449
	1.08	121.2	2679
	1.25	144.2	4154
2	0.69	99.0	810
	0.87	47.7	1383
	1.08	108.2	2457
	1.25	147.9	3840

The results showed that the higher the thickness of the porous layer can deform more for a specific load. Fig. 3 shows the von Mises stress distribution. It was observed that a 2 mm porous layer with a strut thickness of 1.25 mm resulted in reduced deformation. This suggests a higher energy absorption capacity, making it particularly suitable for splint applications under masticatory forces. In this configuration, stresses are efficiently transferred from the solid regions to the porous structure, where the struts play a key role in absorbing mechanical loads. This behaviour aligns with the well-known load-distributing characteristics of the BCC architecture. Designing splints with porous structures internally can enhance stress distribution and absorb occlusal forces more efficiently, reducing the risk of localized

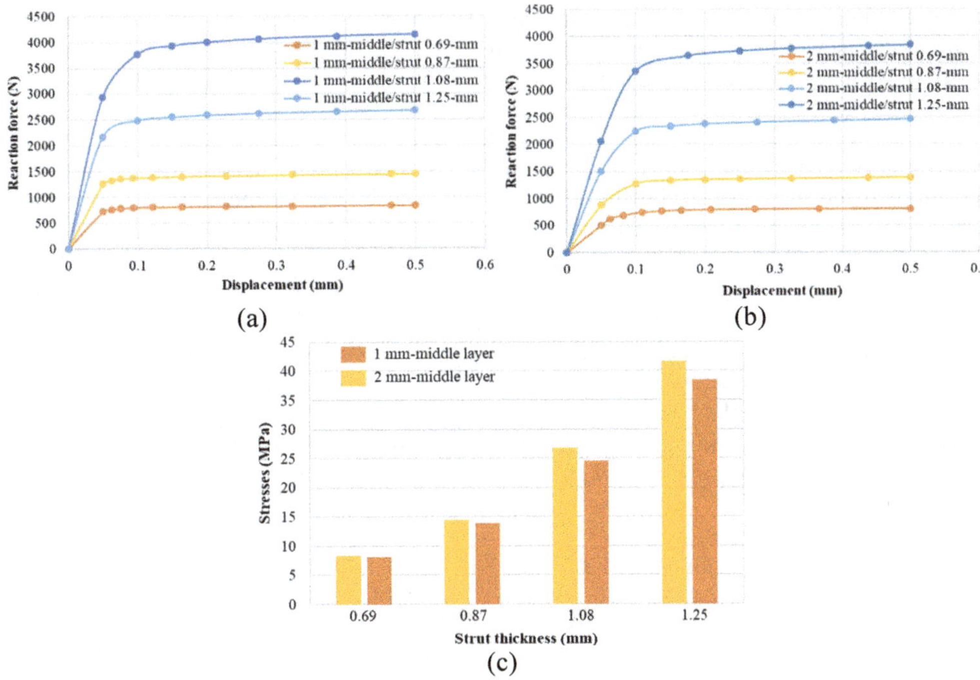

Figure 2: Results of finite element analyses. (a) Compressive behavior and reaction force for 1 mm thick middle layer splint; (b) Compressive behavior and reaction force for 2 mm thick middle layer splint; and (c) von-Mises stresses in different splint models.

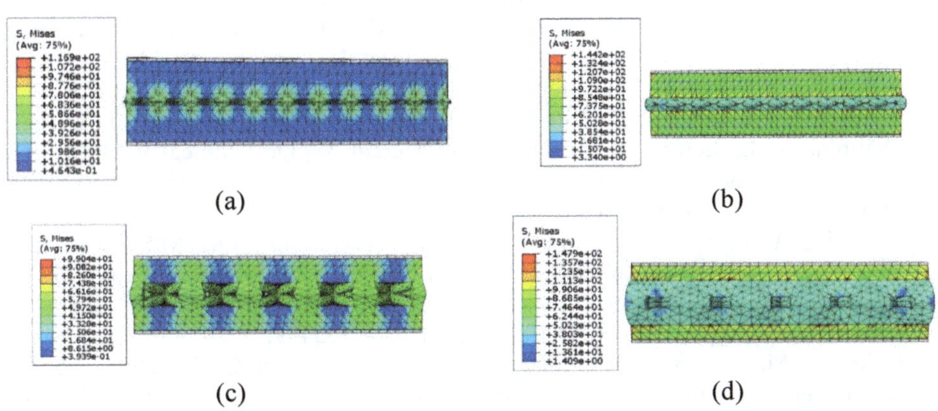

Figure 3: Ghraphical presentation of von Mises stresses and deformation. (a) 1 mm middle layer with 0.69 mm strut thickness; (b) 1 mm middle layer with 1.25 mm strut thickness; (c) 2 mm middle layer with 0.69 mm strut thickness; and (d) 2 mm middle layer with 1.25 mm strut thickness.

pressure points. As a result, patients may experience improved comfort, better protection of underlying tissues, and reduced mechanical stress on dental restorations or implants, making this design especially suitable for long-term use during functional activities such as chewing.

4 CONCLUSION

The results indicated that the porosity of the splints plays a vital role in the stress–strain behaviour when they were subjected to the loads. The higher the porosity, the lower the stresses and the reaction forces. Similarly, increasing the thickness of the porous layer from 1 mm to 2 mm results in a decrease in the stresses and reaction forces for the specific amount of deformation. These results indicate that incorporating porous layers within the splints can effectively distribute reaction forces and mechanical stresses. Thus, introducing the porous layers in the splint may enhance patient comfort with flexible deformation and absorb the stresses during the masticatory forces. Moreover, in the case of dental implants, excess stress transfer to the peri-implant bone can be reduced to prevent bone damage. This design approach can offer several advantages, including enhancing patient comfort, reducing pressure on tissues such as the gingiva and peri-implant contact area.

ACKNOWLEDGEMENT

The authors would like to thank Prince Sultan University for their support.

REFERENCES

[1] Insua, A., Monje, A., Wang, H. & Miron, R.J., Basis of bone metabolism around dental implants during osseointegration and peri-implant bone loss. *J. Biomed. Mater. Res. A*, **105**, pp. 2075–2089, 2017. https://doi.org/10.1002/jbm.a.36060.

[2] Mehboob, H., Ouldyerou, A. & Ijaz, M.F., Biomechanical investigation of patient-specific porous dental implants: A finite element study. *Applied Sciences*, **13**, p. 7097, 2023. https://doi.org/10.3390/app13127097.

[3] Kim, Y., Oh, T., Misch, C.E. & Wang, H., Occlusal considerations in implant therapy: Clinical guidelines with biomechanical rationale. *Clin. Oral Implants Res.*, **16**, pp. 26–35, 2005. https://doi.org/10.1111/j.1600-0501.2004.01067.x.

[4] Häggman-Henrikson, B., Ali, D., Aljamal, M. & Chrcanovic, B.R., Bruxism and dental implants: A systematic review and meta-analysis. *J. Oral Rehabil.*, **51**, pp. 202–217, 2024. https://doi.org/10.1111/joor.13567.

[5] Popescu, A.M., Ionescu, M., Popescu, S.M., Ionescu, A.G., Vlăduțu, D.E., Iacov-Crăițoiu, M.M., Ștefârță, A., Lascu, L.C. & Mercuț, V., Oral clinical and radiological signs of excessive occlusal forces in bruxism. *Diagnostics*, **15**, p. 702, 2025. https://doi.org/10.3390/diagnostics15060702.

[6] Chisini, L.A., Pires, A.L.C., Poletto-Neto, V., Damian, M.F., Luz, M.S., Loomans, B. & Pereira-Cenci, T., Occlusal splint or botulinum toxin-a for jaw muscle pain treatment in probable sleep bruxism: A randomized controlled trial. *J. Dent.*, **151**, 105439, 2024. https://doi.org/10.1016/j.jdent.2024.105439.

[7] Osiewicz, M.A., Werner, A., Roeters, F.J.M. & Kleverlaan, C.J., Effects of occlusal splint therapy on opposing tooth tissues, filling materials and restorations. *J. Oral Rehabil.*, **48**, pp. 1129–1134, 2021. https://doi.org/10.1111/joor.13235.

[8] Nassif, M., Haddad, C., Habli, L. & Zoghby, A., Materials and manufacturing techniques for occlusal splints: A literature review. *J. Oral Rehabil.*, **50**, pp. 1348–1354, 2023. https://doi.org/10.1111/joor.13550.

[9] Hegab, A.F., Youssef, A.H., Al Hameed, H.I.A. & Karam, K.S., MRI-based determination of occlusal splint thickness for temporomandibular joint disk derangement: A randomized controlled clinical trial. *Oral Surg. Oral Med. Oral Pathol. Oral Radiol.*, **125**, pp. 74–87, 2018. https://doi.org/10.1016/j.oooo.2017.09.017.

[10] Ouldyerou, A., Mehboob, H., Merdji, A., Aminallah, L., Mehboob, A. & Mukdadi, O.M., Biomechanical analysis of printable functionally graded material (FGM) dental implants for different bone densities. *Comput. Biol. Med.*, 106111, 2022. https://doi.org/10.1016/j.compbiomed.2022.106111.

[11] Ouldyerou, A., Aminallah, L., Merdji, A., Mehboob, A. & Mehboob, H., Finite element analyses of porous dental implant designs based on 3D printing concept to evaluate biomechanical behaviors of healthy and osteoporotic bones. *Mechanics of Advanced Materials and Structures*, pp. 1–13, 2022. https://doi.org/10.1080/15376494.2022.2053908.

[12] dos Santos Marsico, V., Lehmann, R.B., de Assis Claro, C.A., Amaral, M., Vitti, R.P., Neves, A.C.C. & da Silva Concilio, L.R., Three-dimensional finite element analysis of occlusal splint and implant connection on stress distribution in implant-supported fixed dental prosthesis and peri-implantal bone. *Materials Science and Engineering: C*, **80**, pp. 141–148, 2017. https://doi.org/10.1016/j.msec.2017.05.071.

[13] He, J., Ran, J., Zheng, B., Algahefi, A. & Liu, Y., Finite element analysis of various thickness occlusal stabilization splint therapy on unilateral temporomandibular joint anterior disc displacement without reduction. *American Journal of Orthodontics and Dentofacial Orthopedics*, **161**, pp. e277–e286, 2022. https://doi.org/10.1016/j.ajodo.2021.09.011.

[14] Da Silva, C.A.G., Grossi, M.L., Araldi, J.C. & Corso, L.L., Can hard and/or soft occlusal splints reduce the bite force transmitted to the teeth and temporomandibular joint discs? A finite element method analysis. *CRANIO®*, **41**, pp. 298–305, 2023. https://doi.org/10.1080/08869634.2020.1853464.

SECTION 5
ENVIRONMENTALLY FRIENDLY AND SUSTAINABLE STRUCTURES

CARDBOARD MICROARCHITECTURES FOR THE SUSTAINABILITY OF INTERNAL SPACE REDEVELOPMENT: A LIFE-CYCLE APPROACH THROUGH SCALED PROTOTYPE

DANIELA LADIANA[1] & CHIARA IACOVETTI[2]
[1]University of 'Gabriele D'Annunzio' of Chieti and Pescara, Italy
[2]A3S Progetti Srl, Pescara, Italy

ABSTRACT

The use of cardboard microstructures is an innovative approach to improving the sustainability of interior space renovations. As urban areas face increasing pressure to use space efficiently, cardboard is becoming a versatile and eco-friendly material that can be adapted to meet diverse architectural needs. The objective of this study is to investigate the life cycle assessment potential of cardboard structures, with a particular emphasis on their capacity to mitigate environmental impact and foster adaptability in interior spaces. The utilization of cardboard, instead of traditional materials, facilitates lightweight and modular designs that can be easily reconfigured or disassembled, thus minimizing waste and resource consumption. Using principles of circular economy, these microarchitectures can be designed for reuse, recycling, or composting, thereby reducing landfill contributions and fostering a more sustainable urban environment. In this context, life cycle assessment plays a key role in identifying the environmental performance of cardboard microarchitectures. Through analysing energy consumption, material sourcing, manufacturing processes and end-of-life scenarios, we can identify opportunities for optimization. Moreover, cardboard microarchitectures present distinct prospects for enhancing indoor environments. Their adaptability enables the creation of multifunctional spaces that can adapt to user requirements, thereby promoting flexibility in urban living. Furthermore, incorporating biophilic design principles can enhance occupant well-being by incorporating natural elements within cardboard structures. In conclusion, the integration of cardboard microarchitectures into interior space redevelopment is a promising strategy for sustainability. By focusing on life cycle optimization, architects and designers can create environmentally responsible, adaptable and aesthetically pleasing spaces. This approach not only addresses pressing urban challenges but also contributes to a more sustainable future for architecture.

Keywords: digital manufacturing, microarchitecture, process optimisation, sustainable design, automation, educational experimentation.

1 INTRODUCTION

In an era of radical transformation, the building sector is called upon to respond to changing housing needs. In this context, technological research assumes a strategic role, proposing innovative and sustainable solutions for the definition of new architectural scenarios. In fact, the changing nature of needs, of the services offered and of the ways in which they are used leads us to reflect on the qualities of the built environment, emphasising the impact of the results of transformations on the natural environment and on the well-being of users.

The 2030 Agenda reminds us of the urgency of protecting the planet by implementing sustainable consumption and production models, promoting responsible management of natural resources and concrete actions against climate change [1]. In this context, innovation in technological and environmental design emerges as a key tool to promote changes in the desired direction. Following the 2030 Agenda, which emphasises the commitment to protect the planet through sustainable consumption and production, the sustainable management of its natural resources, and the adoption of urgent measures against climate change, so that it

WIT Transactions on The Built Environment, Vol 216, © 2025 WIT Press
www.witpress.com, ISSN 1743-3509 (on-line)
doi:10.2495/HPSM250091

can sustain the needs of present and future generations, it is clear that the role of technological and environmental design of architecture is a pivotal tool with which it is possible to change the way in which transformation towards a decrease in impacts and consumption of energy and matter occurs [2]. In a century marked by the scarcity of raw materials, the climate emergency, the growing global demand for resources, and where design and production optimisation is asserted as a critical element, a possible strategy of action is the realisation of low environmental impact interventions using light and sustainable materials and the creation of spaces adaptable to different functions. Extremely relevant are not only product technologies, but also new process technologies that, applied to design and realisation, play a crucial role in generating an innovative integrated approach from conception to production, capable of minimising waste and maximising the life cycle of products.

2 BEYOND THE FORM: STATIC VS. DYNAMIC

The rapid evolution of housing needs, combined with the obsolescence of much of the existing building stock and the scarcity of resources for its redevelopment, calls for a paradigm shift in the approach to environmental and technological design. It is no longer a question of introducing static and permanent changes, which risk becoming obsolete before they even degrade, but of developing innovative design solutions. Experiences of experimentation in real contexts conducted by the research group, and the state of the art, would seem to support the use of small architectures, or microarchitectures, for interior space modifications, which, inserted into everyday spaces, can dynamically adapt to the needs of users, offering multiple types of use in both public and private contexts [3]. In line with the principles of the 2030 Agenda, which promotes environmental, social and economic sustainability, these design solutions aim to reduce environmental impact using eco-friendly materials and resource-efficient production processes [4]. At the same time, they favour the creation of flexible and adaptable spaces, capable of responding to the different needs of users over time. The suggested approach is, therefore, to overcome the traditional conception of architecture as a static and immutable entity, embracing a dynamic and sustainable approach, capable of combining technological innovation and environmental responsibility [5]. Cardboard microarchitectures can be easily modified and adapted to different needs, extending their life cycle and reducing the need for new construction. They represent a concrete example of the application of circular economy principles to architecture and design. This approach is based on the creation of closed-loop systems, in which materials are reused and recycled, minimising waste and environmental impact. Cardboard, in fact, is a widely available, recyclable and biodegradable material that can be easily reintegrated into the production cycle and is characterised by its lightness and ease of transport, low environmental impact and low cost [6], [7]. Cardboard microarchitectures can be effectively used in the creation of temporary set-ups for events, exhibitions and trade fairs; to create spaces for education and culture, such as libraries, classrooms and exhibition spaces; flexible and adaptable workspaces that meet the needs of companies and professionals. Indeed, cardboard microarchitectures represent a virtuous example of how the circular economy can be applied to architecture and design, offering innovative, sustainable and low-cost solutions for the creation of flexible and adaptable spaces. Moreover, these devices can dynamically transform the user's perception of the place and generate positive changes in the way that space is inhabited. The social dimension of sustainability highlights the need to develop projects that not only mitigate environmental damage but also improve the well-being of individuals and communities [8]. By thinking of architecture in this way, we provide the user with tools that, in addition to the creative aspect of the architectural object itself, allow them to analyse, know and experience the everyday environment in different ways and with different intensity.

Cardboard microarchitectures, thanks to their modular and flexible nature, allow interior spaces to be reconfigured quickly and easily, adapting them to different user needs. This dynamic approach to architecture encourages the creation of comfortable, stimulating and personalised environments, enhancing the well-being of individuals and communities. As part of the International Design and Building Workshop, held at L'Ecole Nationale d'Architecture de Marrakech (ENAM) from 8 to 18 December 2023 and as part of the didactics of the Technology of Architecture and Design degree workshop, in the Architecture degree course at the University of Chieti and Pescara 'Gabriele D'Annunzio' (Unich), a research and didactic experimentation on the described topic was initiated. The creation of a working group made up of students made it possible to conduct an interesting teaching and research experiment, the contents of which are summarised below.

3 OPTIMIZATION TECHNIQUES AND DIGITAL MODELLING

This study is set in the context of the recent evolution of design and production technologies, in which digital takes on a central and enabling role. In fact, digital technologies now characterise our daily lives and are the undisputed protagonists in the production sector, which is oriented towards a new industrial paradigm. In this scenario, Industry 5.0 is configured as a model in which the massive digitisation of production processes enables the automation of mechanised operations, optimising time and costs. The digital thread connecting design and manufacturing facilitates the customisation of construction elements, opening new frontiers to creativity and efficiency [9]. This provides multiple benefits within the supply chain, including product quality, process efficiency and resource optimisation. Regarding this last aspect, the introduction of digital in the field of building component production could optimise, both in terms of calculation and timing, all those operations that were previously entrusted to humans and not to machines. Indeed, computer graphics research is increasingly concerned with the analysis and high-level processing of geometric objects. By gaining a structural or functional understanding of 3D forms, researchers can tackle medium to high-level design problems for which machine calculations can replace or at least alleviate human production efforts [10]. This new type of process organisation would therefore lead not only to better results in the definition of structure and form, but also to a reduction in the percentage of errors and imperfections in the finished product, thus allowing for less dissipation of the resources used in the production process and the consequent lower production of waste material.

In this context, the development of innovative solutions relating to lightweight, recycled and recyclable materials, as well as the growing interest in prefabrication methodologies for the realisation of easily removable and reusable configurations, have assumed an increasingly important role [11]. Considering the theoretical–practical implications, the experimental work conducted, which had as its object the creation of a multifunctional honeycomb cardboard pavilion prototype, is described below subdivided into consecutive phases, logically ordered according to a well-defined didactic path, characterised by a circular scheme (Fig. 1) aimed at the continuous improvement of the process and the product. In this process, manual dexterity in the practice of drawing and modelling, which begins and concludes the entire process, is central, alongside which the use of computer-based instrumentation, which enriches and implements the empirical phases of the work, is equally important.

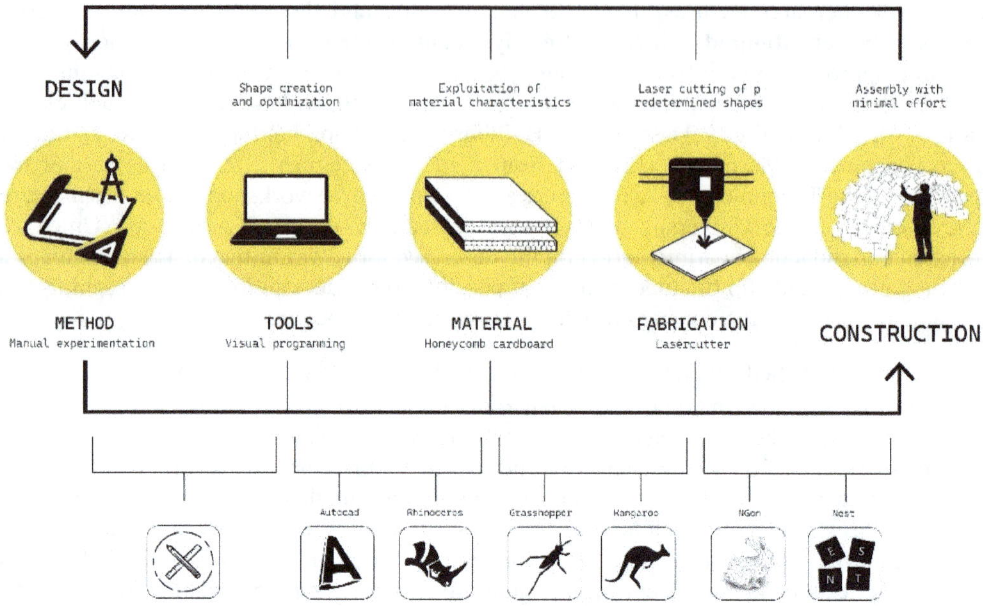

Figure 1: Conceptual map of the project cycle.

3.1 Step 1

The starting point consists of an initial experimentation based on manual drawing, in which the students can try their hand at the elaboration of different proposals for the basic elements that will go on to make up the final structure. The objective of this first phase of work consists in exploring new forms using classical drawing tools, thus promoting the physical and tangible search for innovative architectural solutions, for building elements characterised by different ways of aggregation, flexible and resilient (Fig. 2). During this first step, several proposals of building elements and assembly configurations were formulated, each of which was entirely designed and investigated by means of a hand-built model, to screen each proposed hypothesis and select the most convincing one. Referring specifically to one of these experiments, illustrated here as an example as it was developed for all phases, it is evident how the single piece was reworked several times (Fig. 3):

- The first series of elements, indicated as A-B-C, is characterised by main cuts tangent to the edges of the piece of 2.5 cm depth and two secondary cuts, inclined by 7° with respect to the (central vertical) reference axis, of 1 cm depth. During the assembly phases, however, an excessive rigidity of the parts created through the interlocking of the components just described was noted. Due to the too pronounced depth of the secondary cuts, it was not possible to obtain the correct flexibility of the elements, which, during assembly, were so cohesive that they did not allow the natural curvature of the structure.
- The second series of elements, indicated as A′-B′-C′, is characterised by main cuts tangent to the edges of the piece of 2.5 cm depth and two secondary cuts, inclined by 7° with respect to the (central vertical) reference axis, this time of 0.3 cm depth. Following the modifications made to the model, it was found, during the initial assembly phases,

that the structural elements had poor cohesion at the central cuts, which, being too small, did not stabilise the arch, leading the entire structure to collapse.

- The third and last series of elements, indicated with the initials A″-B″-C″, is characterised by main cuts tangent to the edges of the piece with a depth of 2.5 cm and two secondary cuts, inclined by 7° with respect to the (central vertical) reference axis, increased to 0.5 cm in depth. In view of what has just been described, it is clear how the last modification made to the central cuts proved to be a winner, allowing the structure to be stabilised and at the same time to develop the arc of the circle correctly.

Figure 2: Manual experimentation with pavilions.

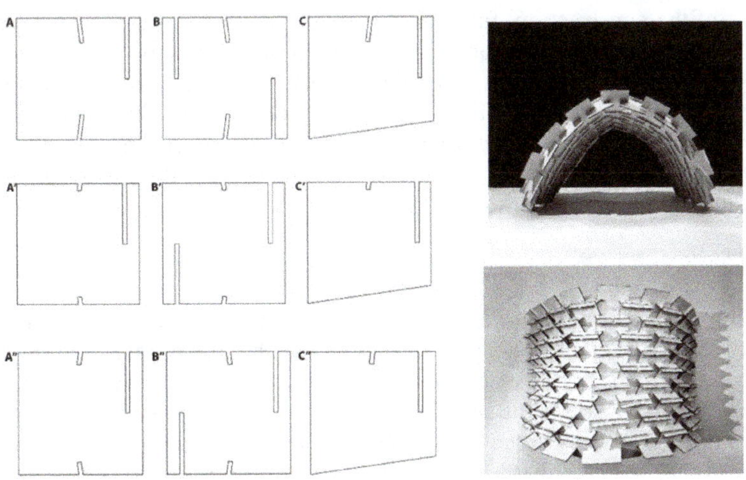

Figure 3: Pavilion first real model.

3.2 Step 2

Following these initial graphic elaborations and manual realisation of models, it was necessary to make use of digital tools capable of returning a very precise description of the object in question. The transition between manual and digital is a fundamentally important moment, in which the empirically conceived architectural element is analysed and tested by means of CAD programmes (AutoCAD) to which the starting information is transferred to obtain an updated and final version. In fact, modelling applied to design is a central tool within the design process as it allows for the exact conception of any architectural composition. The digital model, in fact, is a powerful means of fuelling the creative process and controlling the correctness of the design through the development of the idea in three dimensions. Considering that the design idea is essentially generated within the human mind in a three-dimensional way, it is clear how the use of real or virtual models, during the idea generation process, allows for greater control and a better ability to analyse the project itself, highlighting its criticalities and strengths. By virtue of the organic nature of the process, the design of forms cannot disregard the relationship with the material chosen for realisation: cardboard. A recyclable, ecological and lightweight material with a significant level of toughness and impact resistance, with good thermal and acoustic insulation properties. Considering the properties just described, cardboard is the material of choice for the construction of the final pavilion, which must be economical, ecological and light enough to be assembled and disassembled by anyone quickly and safely, without the need for additional tools.

To maximise the potential for reuse, designers must consider not only deconstruction but also future reconfiguration scenarios [12]. The importance of this last consideration is evident from the final definition of the pavilion's structural components, as, after refining the characteristics of the basic element through digital control tools, it was useful to test an initial assembly of the structure in virtual mode, using 3D modelling programmes. This made it possible to create a spatial vision of the pavilions, understanding potential functional and aesthetic criticalities. To this is added a further methodological deepening, in which parametric variations are made to the previously defined design using plug-ins (Grasshopper, Kangaroo, N-Gon), with the aim of supporting the simple, non-standardised and responsible development of the construction, fully implementing the principles of the complex circular economy (Fig. 4). In this case, the potential of these digital tools for the creation and optimisation of structural elements is exploited to the full, so that even complex organic shapes can be easily achieved (Fig. 5). Indeed, especially in the field of construction, parametric modelling tools allow designers to adjust several underlying parameters to explore a wide range of design outcomes [13], giving them the ability to sift through multiple solutions in less time and more efficiently.

3.3 Step 3

Once all the above criteria were met, the software returned to materiality, with the physical realisation of the shapes thus conceived through the technique of laser cutting first on a 1:3 scale and then, finally 1:1. The innovative charge of the proposed methodology lies both in its mechanical precision and in the speed of production of the desired components, which can be used as soon as the cutting process is finished. In the construction of the 1:1 scale pavilion, a single wall corrugated cardboard panel was employed, comprising a fluted corrugated sheet and two flat cladding sheets. The geometric dimensions of these panels were constrained by the material properties, specifically to prevent the production of excessively elongated

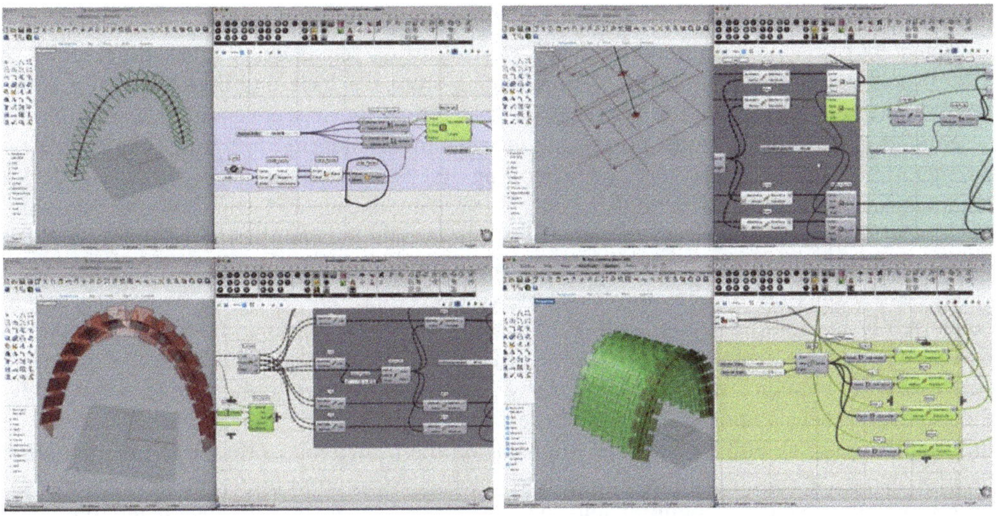

Figure 4: Designing with Rhino and Grasshopper.

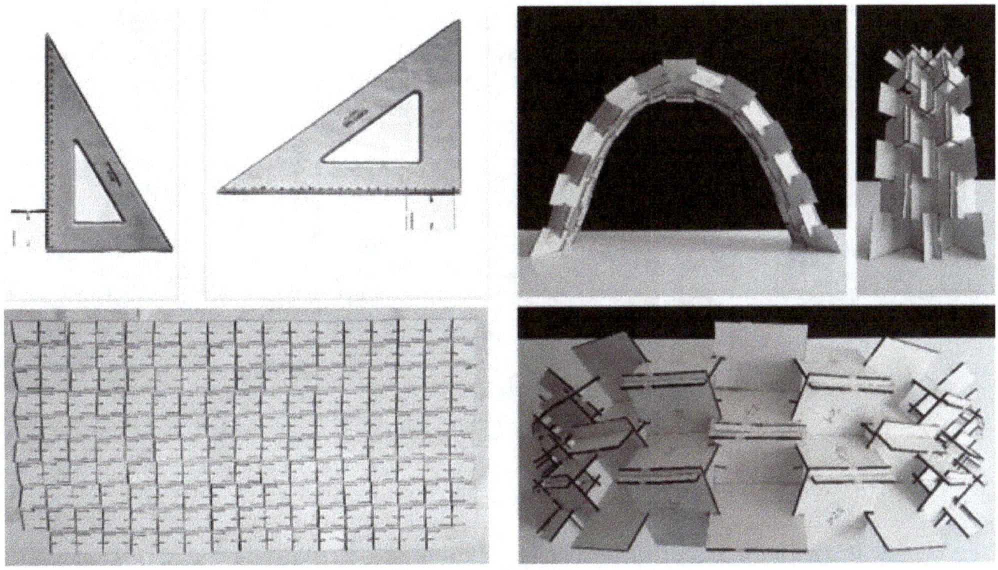

Figure 5: Optimised pavilion prototype.

elements that would compromise stability. An additional constraint arose from the specifications of the laser cutting bed, which limited the number of panels that could be derived from each sheet, in contrast to the smaller prototype models. Although this constraint led to increased fabrication durations, it ultimately yielded significant advantages by expediting the sorting and packing processes, thereby enhancing the efficiency of subsequent construction phases. If the steps just described are carried out effectively, the moment of physical construction of the pavilion prototype can take place simply by nesting the individual pieces made. The assembly procedure, in fact, adhered strictly to the numerical arrangement of the panels, thereby dictating the insertion sequence. A linear trajectory was established to link the components, commencing from the base and ascending in a spiral towards the apex of the shell. It was feasible to segment the sequence and execute multiple assemblies concurrently, whilst preserving the sequential integrity: initiating with a superior enumeration and progressively incorporating the series designated by the inferior enumeration. From a pragmatic perspective, the foundation and supports were initially positioned. Subsequently, in accordance with the sequential numbering of the labels, the edge-stiffened contour components, responsible for load transmission to the supports, were assembled. Adhering to the prescribed assembly order, the structure was constructed concurrently in multiple groups, which then unified along a singular assembly pathway to finalize the components (Fig. 6). The assembly process was executed efficiently: the entire structure was completed in under 15 hours by a cadre of students. In the course of this study, the students were exposed to digital design methodologies and a variety of visual programming tools, with a particular emphasis on the utilization of the physical simulation tool for shape exploration and component processing. Aside from digital modeling, physical prototyping enabled the recognition and analysis of subtle nuances in precision and tolerance, underscoring the critical distinctions between digital and physical models. While the testing phase utilizing scaled models provided a valuable educational experience, greatly appreciated by the students, the hands-on experience of constructing a full-sized structure proved to be even more transformative. The constraints of production, logistics, and material characteristics, exacerbated by the project's scale, emphasized the significance of integrated structural reasoning and construction processes from the initial stages of designing complex architectural entities.

Figure 6: Portion of the archway of the pavilion under construction.

4 CONCLUSIONS

In an era of unprecedented environmental and social challenges, the technological reflection of architecture is called upon to rethink its paradigms and become a promoter of positive change. This study explored the potential of cardboard microarchitectures as an innovative and sustainable response to the redevelopment of interior spaces. Cardboard microarchitectures are constructions capable of responding effectively to the need for functional adaptations and, at the same time, devices capable of transforming the user's perception, generating positive changes in the way space is inhabited. Indeed, human-centred manufacturing requires an emphasis on personalised and 'tailor-made' production, to ensure innovative growth in the manufacturing sector, while responding to the needs of individuals [14].

A growing urban population has, in fact, led to an elevated pressure on consumption of resources in cities. This further leads to significant structural waste and negative environmental impacts in the built environment sector. The need to change and redirect the concept of the transition to a circular built environment is key to achieving a resource-effective and sustainable society. In 2015, the United Nations General Assembly adopted the 2030 Agenda for Sustainable Development with seventeen goals, seven of which are directly related to the implementation of a circular economy. Innovations in the building and construction industry play a significant role in the development of sustainable cities and communities. In this sense, the type of structures designed using cardboard and practices related to parametric design can certainly open new horizons for the sustainable development of architectural environments. Cardboard is a lightweight material, yet it is notably strong and impact resistant. A study by Gribbon and Foerster [15] showed that, although cardboard panels do not have particularly remarkable water absorption properties, their thermal and acoustic performance is impressive compared to other high-performance products on the market. Furthermore, it is possible to design cardboard products with promising insulating properties. From an economic standpoint, cardboard is an affordable material, and its quality can be maintained even in mass production. Mass production and widespread availability are significant advantages for the construction industry, as they help reduce shipping costs. Lastly, the characteristics of cardboard reveal similarities to those of softwood, highlighting both its limitations and strengths.

The educational experimentations conducted demonstrated how cardboard microarchitectures can transform interior spaces, promoting a sense of well-being and connection with the environment. Their modular and reconfigurable nature makes them ideal for a wide range of applications, from temporary fitouts to collaborative workspaces, offering a sustainable and low-cost alternative to traditional constructions. It was described how, through a parametric design approach, it was possible to optimise the performance of these structures, maximising efficiency in the use of materials and minimising environmental impact. The choice of cardboard, a recyclable, low-impact material, proved to be a winner, allowing the creation of lightweight, flexible solutions that can be adapted to the different needs of users. Through this study and in-depth experience, we understand how it is necessary to systematically consider the approach to parametric microarchitecture as an opportunity for didactic experimentation and research, to obtain a new approach to the subject of architecture that can provide flexible answers in an increasingly complex world. Looking to the future, it is essential to continue to invest in the research and development of architectural solutions that integrate the principles of the circular economy and parametric design. Only through a holistic and interdisciplinary approach will it be possible to build a future in which architecture is at the service of man and the planet.

REFERENCES

[1] United Nations General Assembly. A/RES/70/1: Transforming our world: The 2030 Agenda for Sustainable Development. https://sdgs.un.org/2030agenda. Accessed on: 21 Oct. 2015.

[2] Fitas, R., Schaffrath, H.J. & Schabel, S., A Review of optimization for corrugated boards. *Sustainability*, **15**(21), pp. 1–27, 2023. https://doi.org/10.3390/su152115588.

[3] Pilatowicz, G., *Eco-Interiors. A Guide to Environmentally Conscious Interior Design*, John Wiley & Sons: New York, USA, 1995.

[4] Celadyn, M., Environmental activation of inner space components in sustainable interior design. *Sustainability*, **10**(6), pp. 1–12, 2018. https://doi.org/10.3390/su10061945.

[5] Winchip, S.M., *Sustainable Design for Interior Environments*, Fairchild Books: New York, USA, 2011.

[6] Venkatesan, S., Afroz, M., Navaratnam, S. & Gravina, R., Circular-economy-based approach to utilizing cardboard: Sustainable building construction. *Buildings*, **13**(1), pp. 1–18, 2023. https://doi.org/10.3390/buildings13010181.

[7] Klemm, D., Heublein, B., Fink, H.P. & Bohn, A., Cellulose: Fascinating biopolymer and sustainable raw material. *Angewandte Chemie (International ed. in English)*, **44**(22), pp. 3358–3393, 2005.

[8] Omole, C.F.O., Olajiga, O.K. & Olatunde, T.M., Sustainable urban design: A review of eco-friendly building practices and community impact. *Engineering Science and Technology Journal*, **5**, pp. 1020–1030, 2024. https://doi.org/10.51594/estj.v5i3.955.

[9] Mou, T. et al., Bridging the complexity gap in computational heterogeneous catalysis with machine learning. *Nature Catalysis*, **6**(2), pp. 122–136, 2023.

[10] Benes, B., Kasik, D.J., Li, W. & Zhang, H., Computational design and fabrication. *IEEE Computer Graphics and Applications*, **37**(3), pp. 32–33, 2017. https://doi.org/10.1109/MCG.2017.50.

[11] de Gregorio, S., de Berardinis, P. & Rossi, P., MAM: Un modulo abitativo temporaneo personalizzabile e autocostruibile. *Proceedings of the Colloquiate 2020: New Horizons for Sustainable Architecture*, pp. 1137–1153, 2022.

[12] Brancart, S., Paduart, A., Vergauwen, A., Vandervaeren, C., De Laet, L., De Temmerman, N., Transformable structures: Materialising design for change. *International Journal of Design and Nature and Eco Dynamics*, **12**, pp. 357–366, 2017. https://doi.org/10.2495/DNE-V12-N3-357-366.

[13] Yang, Y., Fu, M. & Xing, J., Revolutionizing architecture: The synergy of computational design and digital fabrication. *Applied and Computational Engineering*, **62**, pp. 1–6, 2024. https://doi.org/10.54254/2755-2721/62/20240535.

[14] Lu, Y. et al., Outlook on human-centric manufacturing towards Industry 5.0. *Journal of Manufacturing Systems*, **62**, pp. 612–627, 2022. https://doi.org/10.1016/j.jmsy.2022.02.001.

[15] Gribbon, H. & Foerster, F. Structural engineering and design in paper and cardboard: Approaches and projects. *Cardboard in Architecture*, **7**, pp. 95–117, 2008. https://doi.org/10.3233/978-1-58603-820-5-95.

RETHINKING HEALTHCARE ACCESS: META-DESIGN AND TECHNOLOGICAL INNOVATIONS FOR VULNERABLE POPULATIONS IN THIRD-SECTOR FACILITIES

LISA MENSI* & SILVIA BRUNORO
LEM Technology Laboratory, Architecture Department, University of Ferrara, Italy

ABSTRACT
This paper addresses a critical issue both nationally and internationally: access to healthcare, a fundamental human right, as outlined in Article 25 of the Universal Declaration of Human Rights. This right includes medical care and the broader determinants of health, such as social, economic and physical environments, as well as individual characteristics and behaviours. While these determinants are widely acknowledged in other fields, healthcare design has lagged in addressing holistic well-being within healthcare settings. This issue is particularly significant for individuals who are not eligible for national healthcare systems, such as undocumented immigrants and impoverished populations, who often rely on third-sector organizations for assistance. This study proposes a meta-design framework aimed at improving both the spatial and service design of third-sector organizations that care for marginalized populations. The second part of the analysis explores the integration of appropriate technologies, where interventions often balance routine, temporary, humanitarian and emergency architecture. The goal is to design modular, energy-efficient and flexible clinics that can be assembled and disassembled as needed, while maintaining environmental sustainability – an essential consideration in contemporary construction practices, even for temporary structures. Finally, the implementation of the systems is presented to prove the efficiency and adaptability of the system.
Keywords: third-sector facility, healthcare design, sustainable construction, adaptable construction, design for disassembly.

1 INTRODUCTION

Access to basic healthcare is globally recognized as a human right, as highlighted in Article 25 of the Universal Declaration of Human Rights. This has led more countries to adopt universal healthcare, ensuring medical care for everyone, regardless of income or insurance. However, in some cases, undocumented immigrants are excluded, requiring the third sector to step in to uphold this right and protect public health [1].

In Italy, universal healthcare for all individuals on national territory is guaranteed by Article 32 of the Italian Constitution. The third sector can operate within the National Health Service (NHS), which provides temporary health codes (STP/ENI) to facilitate access to outpatient clinics. Despite this, due to healthcare fatigue, the third sector often remains the primary access point for basic healthcare for these individuals [1]. Barriers to enrolment can arise from policies, systemic issues, and individual factors. In particular, bureaucracy, fear of deportation, communication barriers, and lack of information about the healthcare system play a crucial role in excluding this population [2].

The third sector typically assists in daily life, such as facilitating access to basic medical care, and during emergencies, offering support in the aftermath of crises, including natural disasters, health emergencies, and humanitarian crises.

**ORCID: https://orcid.org/0009-0002-2058-3779*

WIT Transactions on The Built Environment, Vol 216, © 2025 WIT Press
www.witpress.com, ISSN 1743-3509 (on-line)
doi:10.2495/HPSM250101

Similar to everyday life, the quality of temporary shelters significantly impacts both physical and mental health [3], [4]. Current evaluation indicators often fail to assess shelter adequacy fully, overlooking key aspects such as comfort, air quality, and overall well-being [5], [6]. As a result, temporary solutions are frequently used for extended periods, both in daily life and emergencies [7]. Enhancing adaptive design and technological requirements would significantly improve the sustainability of temporary structures [8], enabling a circular economy approach with greater reusability and blurring the lines between temporary and permanent housing [9].

Extensive research has been conducted on hospital architecture design [10], [11]. However, it is equally important to address local healthcare structures. Recent crises highlight the necessity for the proximity healthcare and hospitals to work together effectively in all situations [12].

One of the main challenges in healthcare facility design is its traditional perspective, which often neglects environmental, psychological, and socio-cultural factors in defining health [13]. In particular, the socio-cultural approach views health as a product of social and community structures [14]. A shift in perspective is needed, from rigid, stepwise representations of health outcomes to a more dynamic, multi-faceted model that better captures the complexity of health determinants [15].

The World Health Organization (WHO) has already taken a step in this direction by recognizing that healthcare is influenced by multiple determinants, encompassing not only medical treatments but also social and economic factors. Elements such as poverty and social exclusion are considered key social determinants of health [16] and particularly affect the population target of this study.

The object of this study is to propose a guideline to overlap the gap in the adaptability, reuse, and implementation of third-sector healthcare buildings by proposing an adaptable meta-design applicable to different facilities, aiming to provide a solution for healthcare spaces that need reorganisation or implementation of their spaces, especially in the third-sector. The study intends, also, to provide technological solutions – aiming to practical build the implementation for these facilities – suitable for both long-term and temporary applications in a circular economy perspective.

In the first part of the paper, the meta-design is analysed and explained by examining space characterization, including flow identification, macro-area descriptions, and the categorization of essential spaces required for an outpatient clinic based on accessibility and movement patterns. The meta-design serves as a guideline for organizing these identified spaces according to their function and flexibility, ensuring a replicable and adaptable model suitable for various scenarios, such as new constructions, expansions of existing services, and emergencies.

The second part defines the technological requirements – flexibility, transportability, and adaptability – leading to the development of a technological solution that aligns with the meta-design. The panels are described along with their guidelines.

Finally, the implementation of the system is presented, demonstrating its adaptability across different scenarios and its flexibility in meeting diverse needs.

2 SPACE CATEGORIZATION AND META-DESIGN

The space categorization (Table 2) aims to help reorganize healthcare facilities and serve as a valuable design guideline for broader implementation. Working with existing facilities – each with its own unique needs – requires a flexible meta-design. This approach ensures adaptability to various scenarios while maintaining the efficiency of every service.

The subsequent analyses in this study are grounded in a participatory approach (Table 1) and are informed by the findings emerging from these interviews.

Table 1: Qualitative analysis approach.

Method	No.	Localisation	Data collection		
			Tool	Analysis criteria	Output
Direct survey	10	Emilia-Romagna (6), Marche (1), Piemonte (1), Lazio (1)	Analysis of services and spaces Data collection of access for each structure	Descriptive statistical analysis Comparative analysis among the structures	Services list (Table 2) Macro-areas and flows (2.1/2.2)
Indirect survey	2	Campania (1), Lazio (1)	Analysis of services and spaces	Thematic qualitative analysis	Importance of flexibility (3.1)
Semi-structured interview	15	Emilia-Romagna (11), Marche (1), Piemonte (1), Lazio (2)	Interview guide	Thematic qualitative analysis	Implementation of services and lack of social space in the services (Table 2: 2.1/2.2/3.1)

2.1 Flows

Some areas must be restricted to professionals due to the presence of sensitive materials or confidential information, for this reason also the flows are classified and associated with every space, to regulate the entrance. Five types of movement flow have been identified: F01.1 Public: everyone can access at any time; F01.2 Public after check-in: accessible only to those who have completed the triage process; F01.3 Public under supervision: users are allowed in this area but must always be supervised; F02.1 Private: restricted to staff only; F02.2 Private with public selective access: typically, a private area, but under specific conditions approved by triage or medical staff, users may enter while supervised.

2.2 Macro-areas

In the development of this project, the spaces are divided into four categories, primarily to classify functions, monitor user and professional fluxes, and optimize the strategic layout of services for better efficiency and future adaptability. Additionally, the social area is designed to bridge gaps by creating a new community space focused on the social determinants of health.

2.2.1 Complementary area/A01

The complementary area includes essential services that support the overall healthcare system. It encompasses waiting areas, including a covered external waiting area to protect against weather. Since the service cannot open 24/7, healthcare professionals highlight the need for these shelters, as long queues often form before services open.

2.2.2 Administration area/A02

Directly connected with triage, the administrative area is a staff-restricted access only. However, the administration office is often used by some associations as a reserved space to host users with special needs or to help users with financial support. Ideally, the administration office should also collaborate with local health authorities to activate STP/ENI codes for undocumented migrants, helping to prevent miscommunications that lead the patients to give up on the enrolment and resort to emergency rooms for conditions that could be treated in primary care.

2.2.3 Medical area/A03

The access to the medical area is restricted to those who have completed the triage process. A crucial but often missing feature in outpatient clinics is a dedicated infirmary. The interviewed professionals stress its importance, especially for patients needing frequent medical dressings due to their living conditions.

The peculiarity of the third-sector outpatient clinic is the flexible doctor's office, which hosts a general practitioner but is also equipped to provide other specialized care classified as basic healthcare, such as gynaecology, orthopaedics, and dermatology. Other essential specialist services, which may not always be available in existing structures, include dentistry and psychiatric support/listening centres.

2.2.4 Social area/A04

The interviewed professionals and the literature report that patients often face not only medical challenges but also social issues such as isolation and poor living conditions [13], [14], [16]. To address this, this project developed a new approach incorporating a social area within outpatient clinics. A dedicated daytime space for socializing, combined with a multifunctional room, could host prevention and education programs in collaboration with schools or associations, fostering a more inclusive and effective healthcare system.

2.3 Meta-design development guideline

Once the spaces are classified, a meta-design is proposed to guide the optimal organization and interconnection among spaces (Fig. 1). This configuration includes the essential spaces required to optimize the model and their relationship.

The entrance should lead to a covered waiting area (Fig. 1, Table 2: 01.1.1), which shelters users before triage (Fig. 1, Table 2: 04.1.1). Triage directs users to one of the clinic's three main services (A02/03/04) and grants access to the internal waiting area (Fig. 1, Table 2: 01.1.2) and its services (Fig. 1, Table 2: 01.1.3) for those screened. It also provides access to the infirmary (Fig. 1, Table 2: 02.1).

The administration office (Fig. 1, Table 2: 04.1.2) is the only space accessible to the public under selective access (F02.2), ideally, to offer NHS enrolment support and financial services. Other administrative spaces, like the pharmacy (Fig. 1, Table 2: 04.2.1), storage (Fig. 1, Table 2: 04.2.2/3), and staff changing rooms (Fig. 1, Table 2: 04.3), are restricted to staff.

The medical area should have direct access to the waiting room (Fig. 1, Table 2: 01.1.2) and include various healthcare services, like flexible doctor's offices (Fig. 1, Table 2: 02.2.2), dental offices (Fig. 1, Table 2: 02.2.3), and listening centres (Fig. 1, Table 2: 02.2.4). It is recommended to include two flexible consultation rooms, accommodating both specialist (e.g. gynaecology) and general doctor consultations simultaneously.

The triage should allow access as well to the social area composed of a daily space (Fig. 1, Table 2: 03.1.1) for sociality and a multifunctional room (Fig. 1, Table 2: 03.1.2) for health education, prevention programs, and community collaborations, enhancing inclusivity.

Fig. 1 shows the relationship among the above-mentioned spaces and works as a guideline for design purposes.

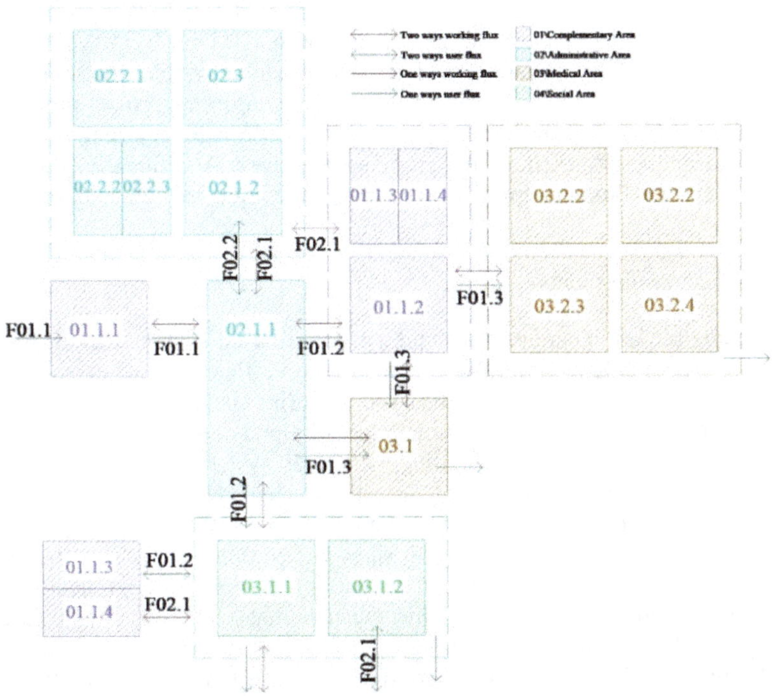

Figure 1: Schematisation of meta-design and relationship (flows) among essential spaces. Purple = A01/complementary area; Blue = A02/administration area; Orange = A03/medical area; Green = A04/social area; Red arrow = professional flow; Green arrow = user flow.

Table 2: Space categorization and allowed accesses.

Category code	Code	Space	Flow
A01: Complementary area	01.1.1	External cover area	F01.1
	01.1.2	Internal waiting area	F01.2
	01.1.3	User toilet	F01.1
	01.1.4	Staff toilet	F02.1
A02: Administration area	02.1.1	Triage	F01.1
	02.1.2	Administrative office	F02.2
	02.2.1	Pharmacy	F02.1
	02.2.2	Clean deposit	F02.1
	02.2.3	Waste deposit	F02.1
	02.3	Staff changing room	F02.1
A03: Medical area	03.1	Nursing	F01.3
	03.2.2	Flexible doctoral office	F01.3
	03.2.3	Dentist	F01.3
	03.2.4	Listening centre/psychiatrist	F01.3
A04: Social area	04.1.1	Daily space	F01.3
	04.1.2	Multifunctional lab	F01.3

3 REQUIREMENTS AND TECHNOLOGICAL DEVELOPMENT

This section presents the technological design of the modular system, focusing on the needs identified in the meta-design and offering a technical solution for project development and implementation. Specifically, these needs were defined through key requirements that the technology must address to achieve the study's objectives. Additionally, the panels were categorized to ensure flexible applications within the meta-design and implementation process.

3.1 Technological requirements

To develop a flexible and adaptable structure, a panel-based system was chosen, meeting three key service requirements: modularity, construction efficiency, and ease of dismantling. This approach aligns with design for disassembly principles and the circular economy, emphasizing sustainability through technological solutions and natural materials that offer environmental benefits while complementing the overall system.

By utilizing a predefined panel catalogue, the system supports customizable layouts tailored to the specific needs of each outpatient facility, which may vary over time. The integration of these panels with building finishes and systems simplifies the construction process, reducing overall build time and promoting a holistic approach to temporary structures with a focus on wellbeing. This ensures that temporary healthcare facilities maintain high-quality standards, reducing the risk of short-term solutions becoming permanent fixtures that lack durability or comfort. Once no longer needed, the structure can be removed, allowing for the restoration of urban space without additional land consumption.

The design also prioritizes the use of natural materials, reinforcing sustainability, and supports a biophilic approach, enhancing energy efficiency and off-grid performance, and ensuring the long-term viability of the building.

3.2 Panels categorization

The panels are designed to be flexibly paired, thus adapting to the needs of various situations. To achieve this while optimizing the production process, a matrix of different solutions has been developed (Table 3), each tailored to meet the main service requirements: modular structure, rapid construction, easy assembly and disassembly, easy maintenance, and well-being over the required period.

The panels are categorized based on the technology used in their design. To streamline both design and production, panels with similar functions typically share the same technological structure and differ only for finishing or additional components.

3.2.1 Standard panels/P01

The standard panels (Fig. 2(a)) are available in two dimensions: 01.1.a (0.6×2.1 m) and 01.1.b (0.6×0.6 m), (Table 3: 01.1). Both dimensions are essential for constructing the system. Specifically, the 01.1.b (0.6×0.6 m) panels are required to achieve the standard height for habitability and are used above the feature panels that do not need a smaller version. Standard panels are utilized when no special features are needed, and they can be used as vertical and horizontal closures.

3.2.2 Systems panels/P02

The system panels (Fig. 2(b)) include one panel combining both electrical and heating systems (Table 3: 02.1), and another panel dedicated solely to the electrical system (Table 3:

Table 3: Panel categorization. a and b define the dimensions applicable to the panel; V, vertical; F, floor; and C, ceiling; specify whether a panel requires slight modifications to connect with others in one of these specific units. When nothing is specified, a standard panel can be used.

Category code	Code	Panel	a	b	V	F	C
P01: Standard panels	01.1	Standard	0.6×2.1	0.6×0.6	–	–	–
P02: System panels	02.1	Electric + Heating	0.6×2.1	–	02.1.1	02.1.2	–
	02.2	Electric	0.6×2.1	–	02.2.1	02.2.2	–
P03: Integrated panels	03.1	Equipment	0.6×2.1	–	–	–	–
	03.2	Mycelium	0.6×2.1	–	–	–	–
	03.3	Biophilic	0.6×2.1	–	–	–	–
P04: Aperture panels	04.1	Door	0.6×2.1	0.6×0.6	–	–	–
	04.2.1	Window: one	0.6×2.1	–	–	–	–
	04.2.2	Window: two	0.6×2.1	–	–	–	–
P05: Structural panels	05.1	Structural	0.6×2.1	0.6×0.6	05.1	05.1	05.1
	05.2	Corner structural	0.6×2.1	0.6×0.6	05.2	05.1	05.1
P06: Hydraulic panels	06.1	Sink	0.6×2.1	–	06.1.1	06.1.2	–
	06.2	WC	0.6×2.1	–	06.2.1	06.2.2	–

02.2). To ensure proper earth connection, a specific configuration is required for floor panels. In particular, the vertical panel (Table 3: 02.1.1) must be connected to the floor panel (Table 3: 02.1.2) to minimize on-site labour and simplify the installation process.

3.2.3 Integrated panels/P03

The integrated panels (Fig. 2(c)) serve different functions, but the proposed technological system remains consistent. It features an internal structure designed to support external loads, such as shelf panels for pharmaceuticals or equipment (Table 3: 03.1). Additionally, the panels incorporate green elements (Table 3: 03.2/3) aimed at creating a biophilic environment for users. One such element is the potential use of mycelium (Table 3: 03.3) in panel construction, which is hypothesized to offer promising capabilities in enhancing air quality [17].

For these panels, no specific adjustments for vertical and horizontal closures are required.

3.2.4 Aperture panels/P04

Due to the specific requirements above the door panel (Fig. 2(d), Table 3: 04.1.a), a dedicated panel (Table 3: 04.1.b) is necessary in place of the standard one (Table 3: 01.1.b). However, for window openings (Fig. 2(e), Table 3: 04.2.1/2), no above panel modification is required, as the standard panels suffice. It is possible to design one or two casement windows.

No specific adjustments for horizontal closures are required.

3.2.5 Structural panels/P05

The structural panels, linear (Fig. 2(h)) and angular (Fig. 2(i)), must be positioned at regular intervals, with no more than three non-structural panels between two structural panels. Each structural panel (Table 3: 05.1/2) must be paired with a smaller version above it (Table 3: 05.1/2.b) and combined with a structural horizontal panel. However, for both vertical and horizontal enclosures, no modifications are required, meaning the same structural panels (Table 3: 05.1 or 05.2) can be used for all the versions.

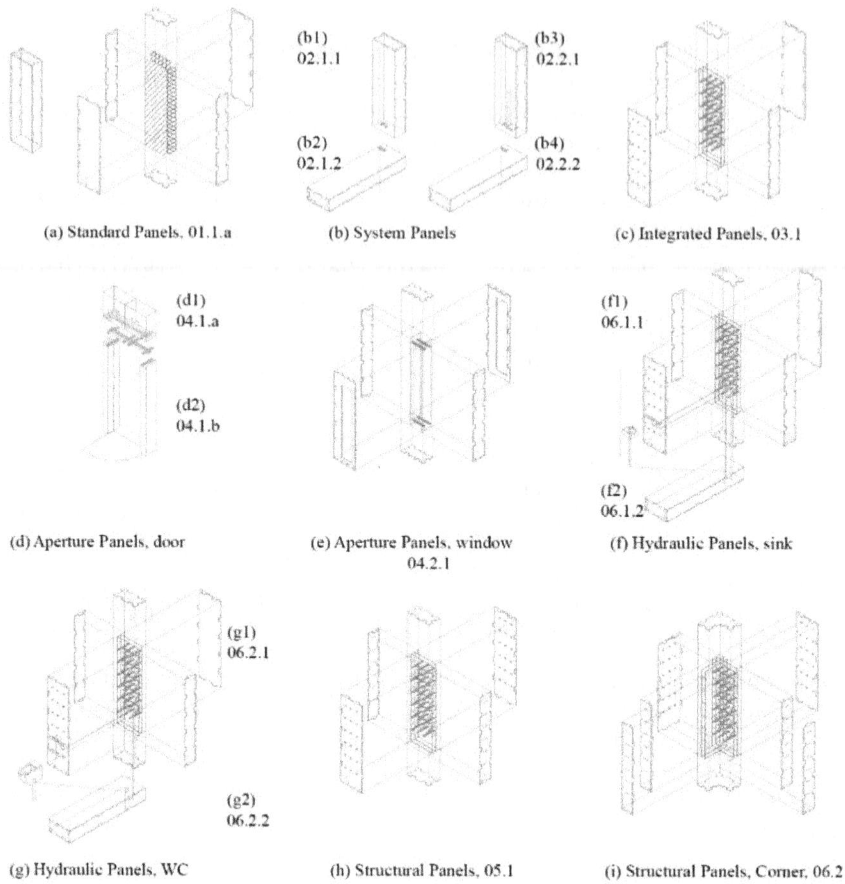

Figure 2: Panel matrix in reference to Table 3. (a) Standard panels; (b) System panels; (c) Integrated panels; (d) Aperture panels, door; (e) Aperture panel, window; (f) Hydraulic panel, sink; (g) Hydraulic panel, WC; (h) Structural panels, linear; and (i) Structural panel, angular.

3.2.6 Hydraulic panels/P06

The hydraulic panels are designed to ensure the proper functioning of bathrooms (Fig. 2(g)) and consultation rooms, each requiring at least one sink (Fig. 2(f)) according to regulations. As happened for the system panels, a specific configuration is needed for the floor panels (Table 3: 06.1.2, 06.2.2) to ensure proper installation and minimize on-site labour. No specific adjustments for vertical and horizontal closures are required.

3.3 Minimal cell and standard designs

The minimal cell was designed to provide a flexible space solution for urgent situations (Fig. 3(a)), offering a standardized approach, which can include all the services indicated in the meta-design. It consists of 6×6 panels, with the first layer made of 0.6×2.1 m panels, and a second layer of overlapping 0.6×0.6 m panels, to meet the minimum room height requirements of 2.70 m.

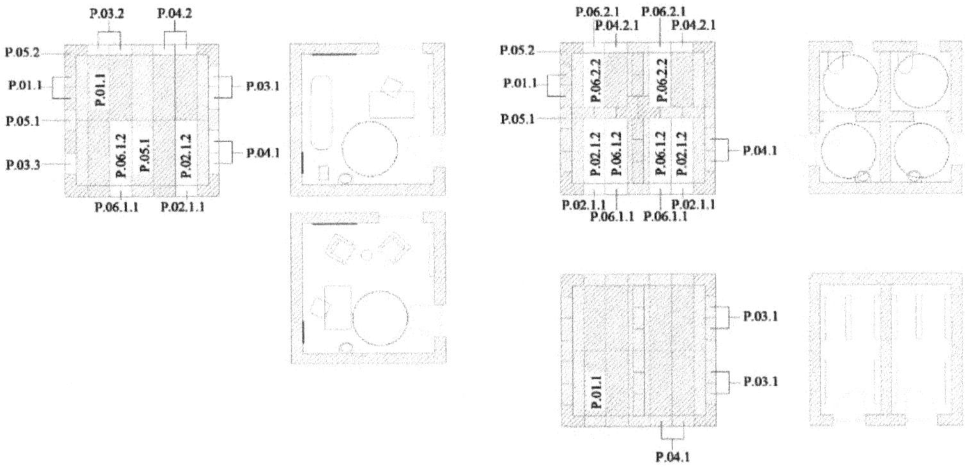

(a) (b)

Figure 3: Minimal cell in standard designs with reference to Table 3. (a) Adaptation of the minimal cell design to categorized spaces: a. up flexible doctor's office, a. down listening centre; and (b) Optimization of the minimal cell design for space efficiency: b. up bathroom, b. down storage areas.

This adaptable design serves as a functional solution, accommodating other essential services such as bathrooms and storage areas (Fig. 3(b)), making it an effective model for emergency and flexible use. The selected panels are a suggestion and can be combined as desired.

4 IMPLEMENTATION OF THE SYSTEM

The system is designed for adaptability, making it suitable not only for third-sector settings but also for a broader range of applications. It can be customized to address varying needs over time, such as fluctuating demands, seasonal changes, or emergencies. Specifically, the outpatient clinic design can be implemented in four different ways.

4.1 Implementation of the meta-design

The meta-design framework allows for flexible implementation, accommodating the integration of individual modules or entire functional areas according to site-specific requirements. This approach is particularly advantageous in new-build scenarios, where it provides scalability and adaptability to dynamic needs. The example shown (Fig. 4) illustrates the addition of an entirely new medical area.

4.1.1 Implementation of the existing facilities

The meta-design has been developed to meet specific needs without requiring all areas to be available for full functionality. By following the provided guidelines for spatial consolidation and flow, existing services can be integrated with missing, substituted, or additional spaces as needed. For example, as shown in Fig. 5, the triage service of an existing facility is enhanced by adding an external canopy, an infirmary, a multifunctional laboratory, and a temporary restroom.

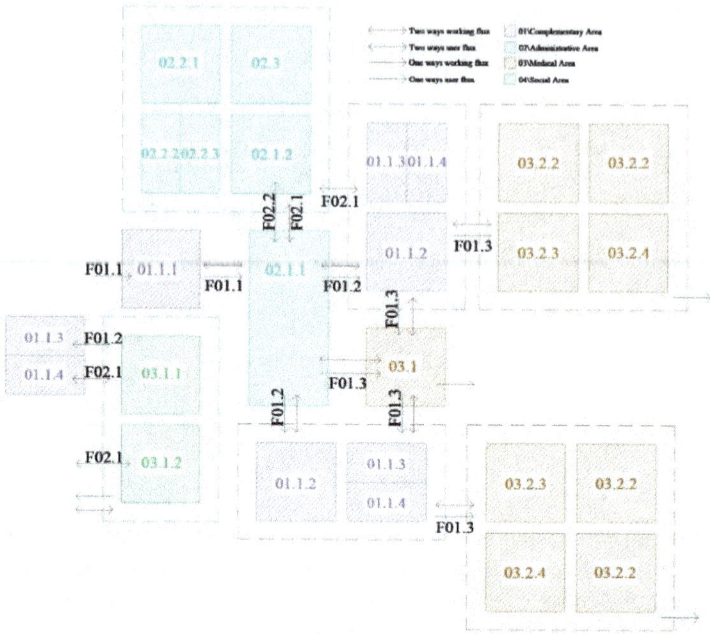

Figure 4: Implementation of the meta-design. This solution illustrates the addition of an entire medical area, based on the hypothesis of increased demand for medical spaces. Similarly, other areas can be expanded using the same approach, following the meta-design approach outlined in Section 2.3.

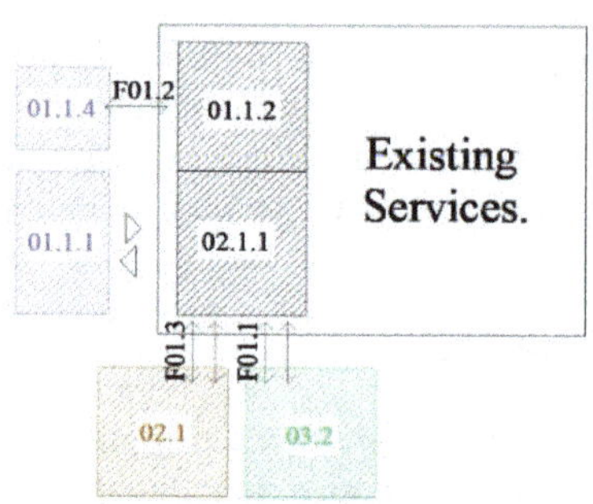

Figure 5: Implementation of existing facilities. In this case, only a few additional spaces are needed to complement the existing services. The same guidelines outlined for the meta-design development (Section 2.3) can be applied to integrate these new spaces.

4.1.2 Implementation of the minimal cell

The spaces are not designed with a fixed layout; rather, the entire system is intended to be customized based on the specific needs and patient flow of the healthcare facility, with the option for temporary implementation. The structure can be easily expanded, for example to accommodate waiting areas or to directly connect a bathroom to a medical consultation room.

This can be achieved in two ways: by adding a minimal additional cell (Fig. 6(a)) or by simply incorporating the necessary panels (Fig. 6(b))

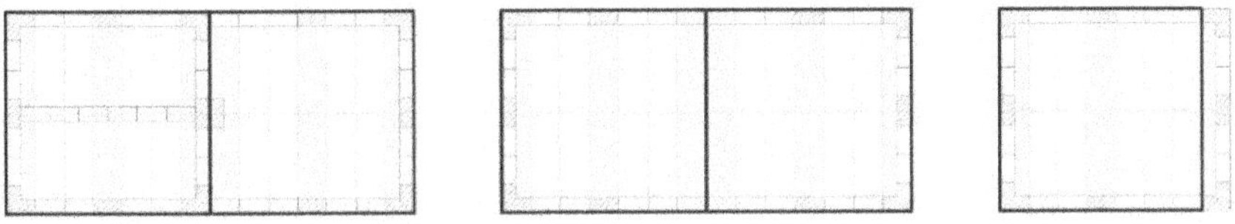

Figure 6: Implementation of the minimal cell. (a) Two examples of the addition of a minimal cell; (b) Addition of single panels.

4.1.3 Implementation of the existing cell
It is also feasible to retrofit an existing structure by replacing panels that require maintenance, or by exchanging standard panels for those equipped with additional functionalities, such as integrated electrical and heating systems (Fig. 7). This approach is particularly relevant when expanding spaces, such as enlarging a waiting area. These adaptabilities make this solution especially suitable for emergency structures, as it allows for seamless modifications in response to evolving operational requirements.

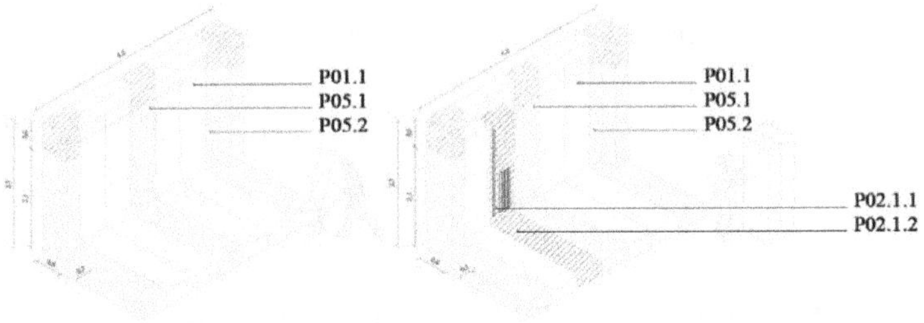

Figure 7: Implementation of the existing cell by replacing one standard panel.

5 DISCUSSION AND CONCLUSION

This study presents a new strategy based on hands-on knowledge for implementing outpatient clinics in the third sector. The proposed system can be used in both everyday services and emergencies to ensure easy access to basic healthcare. It introduces a meta-designed, adaptable solution that meets different needs, along with a technological approach that enables quick assembly, disassembly, and an all-in-one system, eliminating the need for additional on-site construction work. Moreover, the adopted technological strategy ensures easy transportability and zero waste, aligning with the principles of Design for Disassembly and the Circular Economy.

Potential challenges include the cost implications of the project, the available space for the implementation of the existing facilities, and the overall performance of the solution. To address these, further implementation should include testing full-scale panels and evaluating their functionality. Additionally, performance analysis would provide deeper insights into the system's energy efficiency and air quality.

REFERENCES

[1] Quotidiano Sanità, L' evoluzione dei modelli sanitari internazionali a confront. LUISS Business School. quotidianosanita.it/allegati/allegato3331328.pdf. Accessed on: 25 Feb. 2025.

[2] Hacker, K., Anies, M., Folb, B.L. & Zallman, L., Barriers to health care for undocumented immigrants: A literature review. *Risk Manag. Health Policy*, **30**(8), pp. 175–183, 2015.

[3] Asgary, A. & Azimi, N., Choice of emergency shelter: Valuing key attributes of emergency shelters. *Int. J. Disaster Resil. Built Environ.*, **10**(2/3), pp. 130–150, 2019.

[4] Bhui, K. et al., Air quality and mental health: Evidence, challenges, and future directions. *BJPsych Open*, **9**(4), e120, 2023.

[5] Conzatti, A. et al., A review of the impact of shelter design on the health of displaced populations. *Int J Humanitarian Action*, **7**(18), 2022.

[6] Albadra, D., Measurement, and analysis of air quality in temporary shelters on three continents. *Building and Environment*, **185**, 2020.

[7] Askar, R. et al., From temporary to permanent: A circular approach for post-disaster housing reconstruction. *Proceedings of IOP Earth and Environmental Science Conference*, pp. 12–32, 2019.

[8] Johnson, C., Impacts of prefabricated temporary housing after disasters: 1999 earthquakes in Turkey. *Habitat International*, **31**, pp. 36–52, 2007.

[9] Leupen, B., Heijne, R. & Zwol, J. (eds), *Time-Based Architecture*, 010 Publishers: Rotterdam, 2005.

[10] Sari, A. & Jabi, W., Architectural spatial layout design for hospitals: A review. *J. Building Engineering*, **97**, 2024.

[11] Roger, S., A review of the research literature on evidence-based healthcare design. *HERD Spring*, **1**(3), pp. 61–125, 2008.

[12] Setola, N. et al., Hospital responses to COVID-19: Evidence from case studies to support future healthcare design research. *Facilities*, **40**(1/2), pp. 131–145, 2021.

[13] Akram, M., Looking beyond the universal health coverage: Health inequality, medicalism and dehealthism in India. *Public Health Research*, **2**(6), pp. 221–228, 2012.

[14] Advani, M. & Akram, M. *Health Dynamics and Marginalised Communities*, Rawat Publications: Jaipur, pp. 3–26, 2007.

[15] Jayasinghe, S., Social determinants of health inequalities: Towards a theoretical perspective using systems science. *Int. J. for Equity in Health*, **14**(71), 2015.

[16] Chelak, K. & Chakole, S., The role of social determinants of health in promoting health equality: A narrative review. *Cureus*, **15**(1), 2023.

[17] Lee, T. & Choi, J., Mycelium-composite panels for atmospheric particulate matter absorption. *Results in Materials*, **11**, 2021.

SECTION 6
EVOLUTIONARY METHODS IN OPTIMISATION

OPTIMIZING INVERSE KINEMATICS IN THE SCREW THEORY FRAMEWORK WITH QUANTUM HYBRID GENETIC ALGORITHM

SORAYA ZENHARI
Institute for Machine Tools, Stuttgart, Germany

ABSTRACT

On five axis machines, it is necessary to compensate for position independent geometric errors (and position dependent geometric errors) to enhance machining precision. The purpose of this paper is to present an inverse kinematic model which provides clear solutions when executing motion commands according to an inverse kinematic model. A quantum hybrid genetic algorithm is used to determine the parameters of the kinematic model to achieve the most optimal performance possible. A different simulation result validated the proposed approach. This results in an 80% improvement in accuracy compared to conventional genetic algorithms with similar parameters, along with faster convergence. The model is versatile and can be used in a wide range of machine tools, especially those with rotary axes that are not orthogonal. The proposed algorithm can easily be adapted for use in computer-aided design and computer-aided manufacturing systems, depending on the machine's maximum range of motion.

Keywords: screw theory, five axis machine tool, hybrid genetic algorithm, error compensation.

1 INTRODUCTION

The increasing demand for precision manufacturing necessitates advanced machining technologies, especially in industries like aviation, aerospace, and automotive. At the same time, the machine tool industry is focused on improving performance, reducing costs, and minimizing energy consumption to meet the needs of today's economy [1]. Machine tools are prone to geometric errors, which can adversely affect the quality of manufactured components [2]. Among these, position-independent geometric errors (PIGEs) and position-dependent geometric errors (PDGEs) present significant challenges. These errors can arise from imperfections in machine components, thermal deformations, or inaccuracies in the machine's kinematic structure. Compensating for these errors is paramount to ensuring the reliability, quality, and consistency of machined parts. Advanced methods to detect, model, and correct these errors not only improve the precision of the machine tool but also extend its operational efficiency and versatility. To address this challenge, this paper presents an approach that combines screw theory with a quantum hybrid genetic algorithm. The approach is evaluated on real machine data and achieves favourable results. The paper is structured as follows: first, the current state of the art is presented; then, screw theory and the quantum hybrid genetic algorithm are described. Section 5 describes the experiment carried out; Section 6 shows the application of the approach in the form of a simulation. Section 7 then shows the postprocessor. The paper ends with a summary and future perspectives.

2 BACKGROUND AND STATE OF THE ART

In recent years, computer numerical control (CNC) machine tools have become more accurate due to the rapid development of manufacturing engineering. These machines are capable of working quickly and accurately, reducing the need for manual labour and resulting in faster production times. Error compensation allows CNC machines to produce parts with higher precision, reducing waste and improving product quality. It also minimizes the need

WIT Transactions on The Built Environment, Vol 216, © 2025 WIT Press
www.witpress.com, ISSN 1743-3509 (on-line)
doi:10.2495/HPSM250111

for costly rework and increases overall efficiency in the manufacturing process. By enhancing the accuracy of machine operations, companies can achieve more consistent and reliable outputs.

Many researchers have explored various approaches for compensating geometric errors. Using genetic algorithm, Creamer et al. [3] optimized machine tools for effective error compensation. According to Kvrgic et al. [4], geometric errors can be determined in rotational shafts and inverse kinematics in nonorthogonal five axis machines can be analysed. Using a particle swarm algorithm, Chen et al. [5] created a model for compensating geometric errors. A chicken swarm optimization method has been introduced by Fu et al. [6] for optimizing tool directions. The accurate machining of workpieces and cutters requires the use of NC codes generated from cutter location (CL) data. Additionally, each CNC machine requires a specific post-processor.

Based on force-induced, position-dependent errors, Zhang et al. [7] proposed a kinematic chain optimization method for five axis machine tools. By using the homogeneous transformation matrix (HTM) method, they develop a mathematical model to analyse how local deformations affect machining errors. In this process, frequent motion positions are planned, and sensitivity analyses are conducted using static finite element analysis. An alternative kinematic chain's sum and variance indices are used to perform optimization. In comparison with conventional approaches, the method is validated with a five axis vertical machining centre demonstrating a 71% reduction in variance.

By using a postprocessing program in MATLAB, Maletić et al. [8] developed a method for testing a machine's kinematic model. In this study, they presented an inverse kinematic model consisting of five equations and then translated the CL data into a G-code program for five axis machining. Further, they developed software for developing kinematic equations for a multi-axis post-processor and implemented a mathematical model using elementary transformational matrices. Post-processing software successfully converted CL data into G-code for the machining of concave workpiece [9].

Using an iterative control vector parametrization method, Ma et al. [10] recommend optimal federate planning for five axis CNC machines with geometric and kinematic constraints like velocity, acceleration, and jerk. An analysis based on CAD-generated virtual models of the main components of an eight-axis machine tool was conducted by Ullah et al. [11]. This study focused on optimizing the design for better machine performance and structural integrity. The lattice optimization resulted in a 64% reduction in energy consumption. A solution algorithm based on least squares and Particle Swarm Optimization PSO was proposed by Yan et al. [12], considering the error quantity function of each tool position. Zhang et al. [13] focused on determining and compensating for geometric errors in five axis machine tools to improve the quality of the products they produced. Inverse kinematics is used to calculate errors, and volumetric. In another study, Gao et al. [14] proposed an optimization method for the setup of workpieces, which took into account the need to avoid singularities and to ensure that the five axis system was stable kinematically. The solutions to ill-conditioning in the Jacobian matrix of five axis inverse kinematics were discussed by Lu et al. [15]. As part of another study, Dening et al. [16] used the Jacobian condition number to limit tool orientation to prevent singularities.

The combination of geometric error compensation and optimization processes has been studied relatively little. By applying improved genetic algorithms, Wei et al. [17] developed a novel optimization method for obtaining the NC code for geometric error compensation, which used the complete position deviation as the fitness function.

A quantum hybrid genetic algorithm (HQGA) can provide a promising solution for CNC machine tool error compensation and optimization due to its ability to combine quantum computing with classical genetic algorithms, thereby increasing the efficiency and capability of optimization. Hybrid methods excel at solving complex, NP-hard combinatorial problems typically encountered in manufacturing, resulting in improved accuracy and adaptability in dynamic machining environments. With HQGA, unlike traditional genetic algorithms which may become trapped in local optima, and pure quantum methods that are limited by current hardware constraints, HQGA provides an effective framework for addressing the high-dimensional, nonlinear challenges associated with CNC machining, making it the optimal choice for modern manufacturing operations.

3 SCREW THEORY

Screw theory models the relative position and orientation of each axis in a five axis machine tool, offering a clear understanding of its three-dimensional motion. By incorporating both rotational and translational movements, it provides a mathematical framework to describe the relative motion between the tool and the workpiece, enabling a more intuitive approach to kinematic analysis. To enhance the accuracy and reliability of the machine tool, error models must be applied to calculate discrepancies between the tool tip and the workpiece. An accurate representation of a five axis machine tool requires consideration of both the workpiece and the toolchain. Assuming that the machine tool moves by a certain amount (x, y, z, θ_A, θ_C), then the updated coordinates of the workpiece and the tool tip in the machine coordinate system (MCS) can be assessed using screw motions $e^{\hat{\xi}\cdot\theta}$ with twists ξ, described as [18]:

$$g_{bw} = e^{\hat{\xi}_X\cdot(-x)}.\, e^{\hat{\xi}_A\cdot(-\theta_A)}.\, e^{\hat{\xi}_C\cdot(-\theta_C)}.\, g_{bw}(0) \tag{1}$$

$$g_{bt} = e^{\hat{\xi}_Y\cdot(y)}.\, e^{\hat{\xi}_Z\cdot(z)}.\, g_{bt}(0) \tag{2}$$

where $g_{bw}(0)$ and $g_{bt}(0)$ are initial 4×4 transformation matrices of workpiece and tool tip relative to measuring coordinate system. Regarding the forward kinematics model, five axis machine tool was utilized for the experiment, its transformation matrix can be expressed as follows:

$$g_{bw}(0) = \begin{bmatrix} 1 & 0 & 0 & 0 \\ 0 & 1 & 0 & 0 \\ 0 & 0 & 1 & -z0 \\ 0 & 0 & 0 & 1 \end{bmatrix} \tag{3}$$

$$g_{wt} = g_{wb}\cdot g_{bt} = (g_{bw})^{-1}\cdot g_{bt} = g_{bw}^{-1}(0).\, e^{\hat{\xi}_C\cdot(\theta_C)}.\, e^{\hat{\xi}_A\cdot(\theta_A)}.\, e^{\hat{\xi}_X\cdot(x)} e^{\hat{\xi}_Y\cdot(y)}.\, e^{\hat{\xi}_Z\cdot(z)}.\, g_{bt}(0) \tag{4}$$

These relations can be used to drive mathematic equations for inverse kinematics. In the rotary axes, screw motion is defined as follows [19]:

$$\xi = \begin{bmatrix} v \\ \omega \end{bmatrix}, v = q \times \omega \tag{5}$$

where ω is the unit vector in the positive direction of the rotary axis, and q is any point on the axis expressed in the machine coordinate system (MCS). In the rotary axes, screw motion is defined as follows

$$e^{\hat{\xi}\cdot\theta} = \begin{bmatrix} e^{\hat{\omega}\theta} & (I_{3\times3} - e^{\hat{\omega}\theta})(\omega \times v) + \omega\omega^T v\theta \\ 0_{1\times3} & 1 \end{bmatrix} \tag{6}$$

where θ is the angular rotation of the axis and $e^{\hat{\omega}\theta}$ can be expanded as:

$$e^{\hat{\omega}\theta} = I_{3\times3} + \hat{\omega}.\sin\theta + \hat{\omega}^2.(1-\cos\theta) \tag{7}$$

Error twists are used to model the geometric errors between the tool tip and workpiece and will be applied as q_d and ω_d into the equations. The ideal path that the machine should trace will differ slightly from the actual path, as shown in Fig. 1.

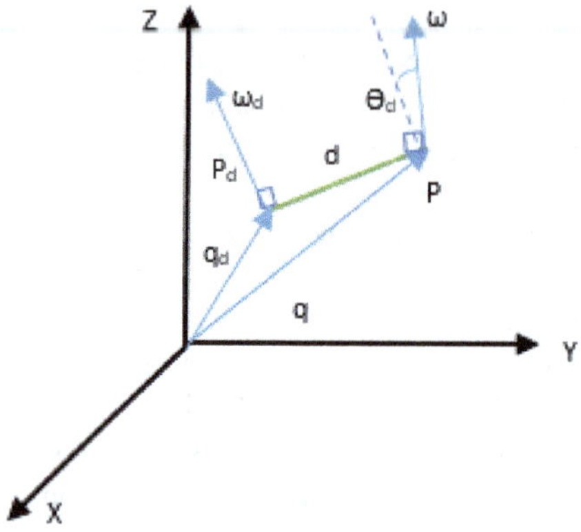

Figure 1: Error twist.

These deviations can lead to inefficiencies that could compromise the quality of the final product or process. As mentioned before the geometric errors of a machine tool can be categorized PDGEs and PIGEs. The five axis machine tool under study suffers from 41 errors. These errors can be detected using metrology techniques and devices such as laser interferometers and double ball bar devices [19]. To address these errors, several measurements were performed on the machine's structure to accurately identify and quantify them. Considering eqns (5)–(7), incorporating the error into the equations for a revolute joint, we can rewrite the equations as follows [19]:

$$\omega_e = \frac{\omega_d \times \omega}{\sin\theta_e}, \; h_e = \frac{d}{A_e} = \frac{|q-q_d|}{\theta_e} \tag{8}$$

$$v_e = \frac{q\times(\omega_d\times\omega)}{\sin\theta_e} + h_e\frac{\omega_d\times\omega}{\sin\theta_e} = \frac{q_d\times q}{d} + \frac{q-q_d}{\theta_e} \tag{9}$$

and for prismatic joints, where ω = 0:

$$\xi_e = \begin{bmatrix} \frac{q-q_d}{d} \\ 0 \end{bmatrix} \tag{10}$$

To correct geometric errors, the axis errors are measured and modelled as functions of position in the machine coordinate system. These errors are then translated to the tool's orientation and tip position using the forward kinematic model of the machine in the part coordinate system. This step considers the impact of axis errors on the tool's final position and orientation during machining. To address these errors, the tool position and orientation errors are converted into corrections for the machine's drive components using the inverse kinematic model in the part coordinate system. This ensures that the tool's movement is adjusted to counteract the measured errors, thereby enhancing machining accuracy.

4 QUANTUM HYBRID GENETIC ALGORITHM

The Genetic Algorithm (GA) is a metaheuristic inspired by natural selection, first introduced by John Holland and later expanded by Goldberg [20]. It employs evolutionary principles such as selection, mutation, and crossover to solve optimization and search problems. GAs are widely used in machine optimization to reduce geometric errors by optimizing machine tool movements, thereby increasing precision and product quality [21]. The algorithm begins with a population of potential solutions, referred to as chromosomes, which are evaluated based on a fitness function reflecting their performance relative to the optimization objective. The process iteratively selects the best solutions to form new generations, gradually improving the population until an optimal solution is found [22].

In the basic GA, the process begins by generating an initial population with diverse solutions. Each chromosome is evaluated using a fitness function, and the fittest individuals are selected as parents. Crossover combines genetic information from two parents to create offspring, while mutation introduces random changes to maintain diversity and prevent premature convergence. These steps population generation, fitness evaluation, selection, crossover, and mutation are repeated until the termination criteria are met, such as achieving a target fitness or reaching a set number of generations. While the basic GA is effective for many optimization tasks, it can struggle with premature convergence or fail to find globally optimal solutions in complex problems. To address these challenges, the Quantum Hybrid Genetic Algorithm (QHGA) incorporates two additional steps: The Newton–Raphson optimization and a quantum-inspired crossover phase, providing enhanced performance and robustness. The QHGA process begins like the basic GA, with the generation of an initial population, fitness evaluation, and parent selection. However, before performing the crossover phase, the algorithm applies the Newton–Raphson optimization method to the top-performing chromosomes. This gradient-based method refines the parameters of selected chromosomes by leveraging the cost function's gradient, enabling more accurate solutions and faster convergence. After this refinement, the crossover phase proceeds as in the basic GA, creating offspring by combining the genetic material of two selected parents. An additional innovation in the QHGA is the introduction of the quantum-inspired crossover phase, which addresses the issue of stagnation. If no improvement is observed in the fitness value according to threshold number, the algorithm detects stagnation and triggers the quantum-inspired crossover step. This phase introduces novel genetic combinations by leveraging probabilistic techniques inspired by quantum mechanics, helping the population escape local optima and continue improving. This mechanism significantly enhances the algorithm's ability to explore diverse regions of the solution space. In quantum-inspired crossover, a matrix called R is used to generate a child from two parents. This matrix will be added into the calculation as follows:

$$x_{\text{Child}} = R\frac{x_1+x_2}{2} \tag{11}$$

where $x_1$and x_2 were considered as selected parents and R is the proposed matrix. It is worth mentioning that R plays an essential role in introducing diversity and effectively exploring the solution space. Resulting in improved performance in trajectory optimization.

Finally, mutation is applied to introduce random alterations in the offspring, ensuring the population maintains genetic diversity and adaptability. The process continues iteratively, repeating the steps of Newton–Raphson optimization, crossover, quantum-inspired crossover (if necessary), and mutation until the termination criteria are met. By combining the Genetic Algorithm (GA) with the Newton–Raphson method and quantum-inspired crossover, the QHGA achieves convergence, prevents premature stagnation, and enhances solution diversity. This hybrid approach makes the QHGA particularly effective for solving complex optimization problems that require both precision and adaptability. The proposed algorithm is illustrated in Fig. 2.

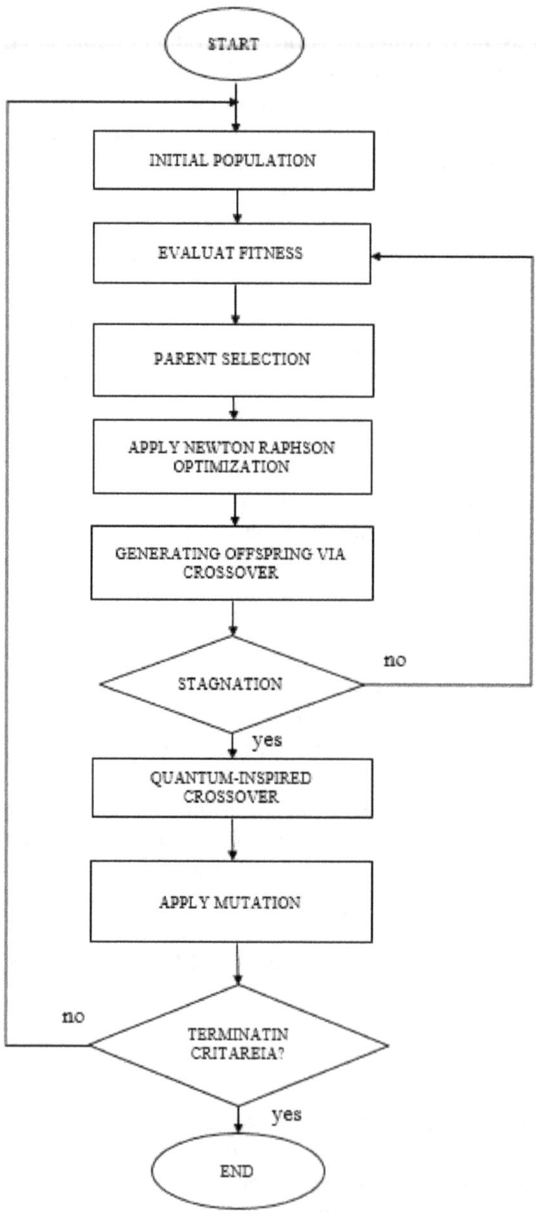

Figure 2: Proposed algorithm flow chart.

5 EXPERIMENT

The kinematic model of a five-axis machine tool consists of a combination of prismatic and revolute joints, enabling precise and flexible machining operations. Typically, such machines feature separate kinematic chains for the tool and workpiece, allowing for multi-axis motion. The tool's movement is often governed by a set of prismatic joints, while the workpiece is manipulated using a combination of prismatic and rotational joints. This structure provides extensive motion capabilities necessary for complex machining tasks. To analyse the inherent accuracy of multi-axis machines and quantify geometric errors, high-precision measurement devices such as laser trackers and double ball bar systems are commonly used. These tools facilitate the evaluation of linear and circular positioning accuracy by performing systematic tests across multiple planes. Circular tests typically assess motion precision in various planes, while linear measurements capture positional deviations along individual axes at different feed rates. By integrating these measurement techniques, it is possible to quantify axis-specific errors and develop compensation strategies to enhance machine accuracy. Geometric errors in multi-axis machines can be systematically analysed and modelled using screw theory. This framework represents both translational and rotational deviations as unified error twists, which integrate linear and angular errors into a single mathematical representation (see eqns (8)–(10)). By expressing these errors in a global coordinate system, screw theory eliminates the need for multiple local reference frames, simplifying error modelling and compensation. Each axis error can be expressed in terms of an exponential screw displacement, allowing for efficient computation of position and orientation deviations (see eqn (8)). The total error in tool positioning can be determined by summing the contributions of individual error components, including positional deviations along linear axes and angular misalignments in rotary joints. Using the product of exponentials (PoE) method, the overall transformation incorporating these errors is obtained, providing a structured approach to error compensation. Additionally, screw theory facilitates the efficient mapping of individual error components, such as those listed in Table 1 (e.g., δ_{xx}, δ_{yy} for translational axes and δ_{aa}, δ_{cc} for rotary axes), to the overall volumetric error. This enables a clear understanding of how specific errors impact the machine's accuracy. By systematically quantifying these deviations, error compensation strategies can be implemented to refine machine performance.

Table 1: Geometric errors (PDGEs and PIGEs) [19].

X-axis	Y-axis	Z-axis	Squareness error	A-axis	C-axis	PIGES of R rotary axes
δ_{xx}	δ_{yy}	δ_{zz}	S_{xy}	δ_{xa}	δ_{xc}	δ_{xoc} S_{boa}
δ_{yx}	δ_{xy}	δ_{xz}	S_{yz}	δ_{ya}	δ_{yc}	δ_{yoc} S_{coa}
δ_{zx}	δ_{zy}	δ_{yz}	S_{xz}	δ_{za}	δ_{zc}	δ_{zoa} S_{aoc}
ε_{xx}	ε_{xy}	ε_{xz}		ε_{xa}	ε_{xc}	δ_{zoa} S_{boc} (ISO 230-7)
ε_{yx}	ε_{yy}	ε_{yz}		ε_{ya}	ε_{yc}	
ε_{zx}	ε_{zy}	ε_{zz}		ε_{za}	ε_{zc}	

Once errors are measured and modelled as functions of position within the machine's coordinate system, a correction algorithm can be implemented to adjust machine movements accordingly. By continuously monitoring and compensating for deviations, this approach improves machining accuracy and consistency, ultimately enhancing the precision and reliability of multi-axis manufacturing processes.

6 SIMULATION

Further investigation is required to evaluate the efficiency of the proposed algorithm. The trajectory of the tool in the work piece's local frame is considered as follows:

$$x_{ref} = \sin(t) \tag{12}$$

$$y_{ref} = \cos(t) \tag{13}$$

$$z_{ref} = 0 \tag{14}$$

The objective is to find an optimal parameter set $[x_1, x_2, x_3, \theta_4, \theta_5]$ that minimize the cost function. According to proposed algorithm (see Fig. 3) after iteration 20 the results convergence to optimal values, considering home position of the machine tool optimized parameters for a machine tool's positioning can be found as: $x_1 = 4.07$ µm, $x_2 = -1.32$ µm, $x_3 = 5.62$ µm.

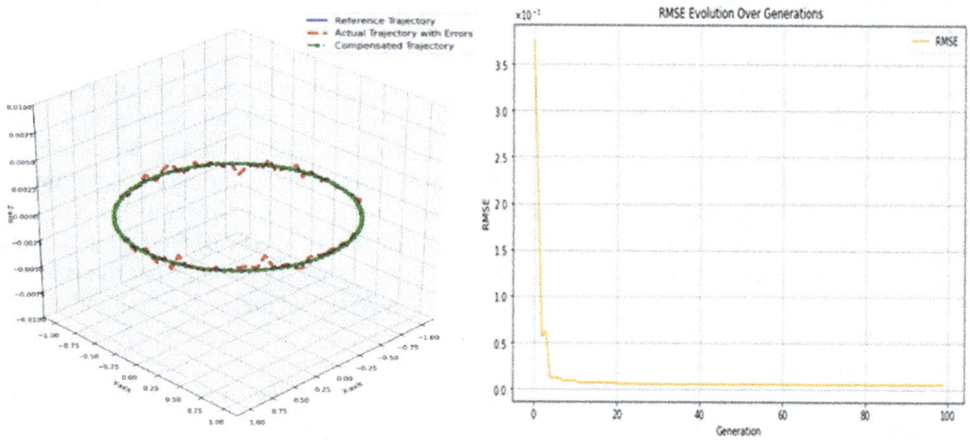

(a) (b)

Figure 3: (a) Reference and actual paths; and (b) RMSE error of the proposed algorithm.

In CNC machining, even small misalignments between the tool and the workpiece can lead to poor surface finishes, dimensional errors, or incorrect part geometries. To maintain high part quality, precise adjustments to the tool's position are necessary. The transformation matrix plays a crucial role in modelling and correcting these displacements by incorporating both translation and rotational components for accurate machining. Operators adjust the matrix parameters to compensate for machine errors. These transformations, typically expressed through rotation matrices and translation vectors, account for both linear and angular movements. The matrix is derived by identifying optimal parameters through empirical measurement, calibration, and simulation, using methods like matrix exponentiation and Euler angles. These adjustments ensure the tool's position is precise during operations. The three key parameters, expressed in micrometres, guide the adjustments required for optimal machine performance. Based on these, the transformation matrix for the tool relative to the workpiece is computed as follows [23], [24]:

$$g_{bt} = e^{\xi_z x1} e^{\xi_y x2} g_{bt0} = \begin{bmatrix} 1 & 0 & 0 & x_0 \\ 0 & 0 & -1 & -x_2 \\ 0 & 1 & 0 & x_1 \\ 0 & 0 & 0 & 1 \end{bmatrix} \tag{15}$$

$$g_{bw} = e^{\xi_1 x3} e^{\xi_{\theta_y} \theta_4} e^{\xi_{\theta_z} \theta_5} g_{bw0}$$

$$= \begin{bmatrix} \cos(\theta_4)\cos(\theta_5) & -\cos(\theta_4)\sin(\theta_5) & -\sin(\theta_4) & 0 \\ \sin(\theta_5) & \cos(\theta_5) & 0 & -y_0 \\ \sin(\theta_4)\cos(\theta_5) & -\sin(\theta_4)\sin(\theta_5) & \cos(\theta_4) & x_3 \\ 0 & 0 & 0 & 1 \end{bmatrix} \tag{16}$$

Fig. 3(a) shows the ideal path, the actual path, and the path with compensation and (b) illustrating the algorithm's efficiency per iteration. The algorithm's efficiency per iteration demonstrates a significant improvement, indicating that the compensation method effectively enhances the overall performance. Consequently, the proposed approach not only refines the path but also optimizes the computational process, ensuring a more efficient and accurate outcome.

7 POST-PROCESSOR

A post-processor was designed based on the concepts to convert data from screw theory (see Fig. 4). The interface is divided into three steps:

1. General: Machine information, such as CAD data provided in the library, is selected in this step.
2. Process and environment: In this step, the volumetric error of the machine tool is uploaded. Based on the forward kinematics framework derived from screw theory, deviations in three dimensions are calculated.
3. Detail design: Users can evaluate and optimize parameters to reduce the corresponding errors. Various optimization methods can be employed in this step. In this paper, we focus on quantum hybrid genetic algorithms, which can provide accurate results.

Based on the designed postprocessor, engineers can easily evaluate the results without delving into complex calculations, allowing them to quickly assess simulation outcomes.

8 CONCLUSION AND OUTLOOK

This research introduces a novel hybrid algorithm for solving inverse kinematics problems, achieving an 80% improvement in proficiency over traditional genetic algorithms with similar parameters. The algorithm effectively addresses stagnation caused by singularities and demonstrates rapid convergence, finding solutions within just a few generations. Moreover, it can determine intermediate configurations for the end effector, enhancing its adaptability in complex scenarios. To improve usability, a user-friendly post-processing step was implemented, making the algorithm more accessible for practical applications. Future work will focus on expanding its adaptability to various machine tools and robotic systems across a broader range of scenarios. Additionally, incorporating Kalman filters to update feedback signals will enable real-time error compensation, enhancing precision during operation. Further enhancements include developing a digital twin architecture, which is expected to improve real-time accuracy and facilitate predictive maintenance, thereby increasing system reliability in dynamic environments. Another key advancement involves

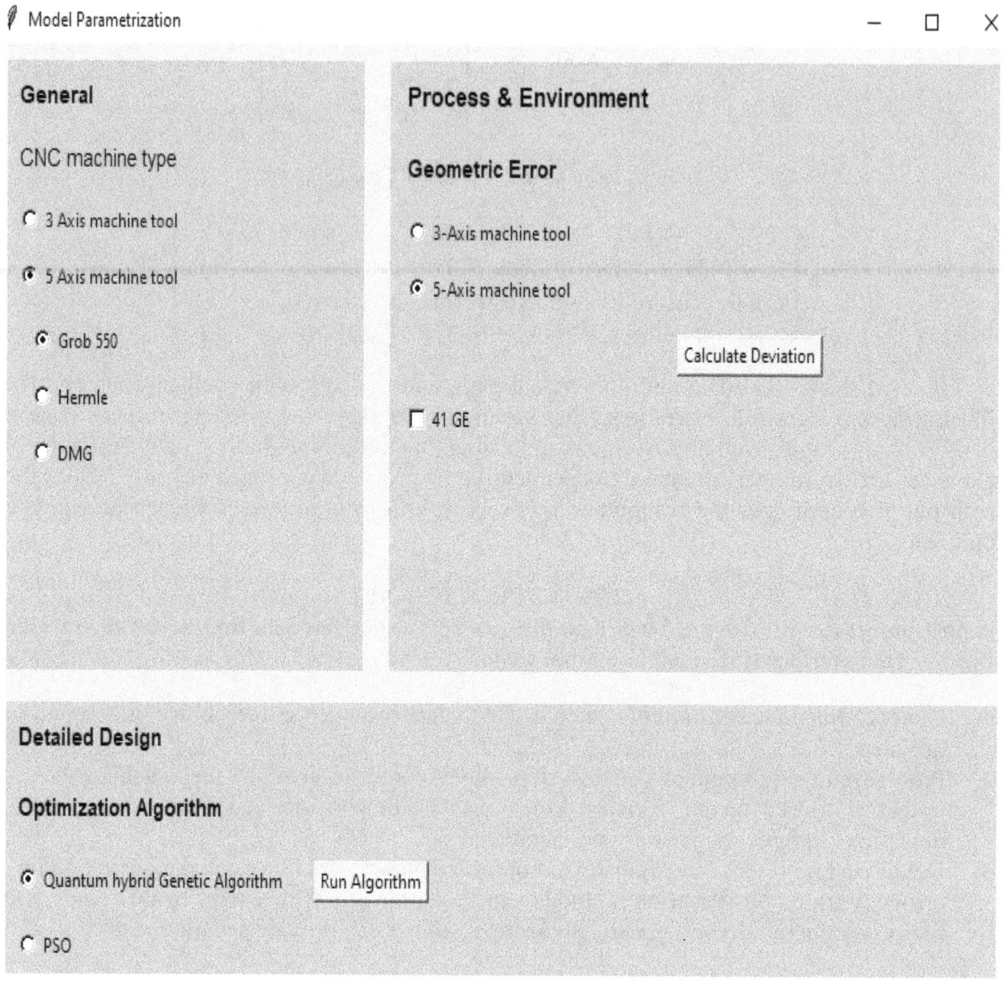

Figure 4: Dialog of the proposed five axis postprocessor.

refining the machine interface's G-code to incorporate error compensation, allowing for more precise control, reduced errors, and increased overall efficiency. Finally, integrating machine learning techniques for continuous optimization could lead to more intelligent and autonomous robotic systems, expanding their applications across industries. These advancements will contribute to greater reliability, adaptability, and efficiency in robotic and CNC-based machining environments.

REFERENCES

[1] Wang, H., Ran, Y., Zhang, S. & Li, Y., Coupling and decoupling measurement method of complete geometric errors for multi-axis machine tools. *Appl. Sci.*, **10**(6), 2020. https://doi.org/10.3390/app10062164.

[2] Zhang, Z., Jiang, F., Luo, M., Wu, B., Zhang, D. & Tang, K., Geometric error measuring, modeling, and compensation for CNC machine tools: A review. *Chinese J. Aeronaut.*, **March**, 2023. https://doi.org/10.1016/j.cja.2023.02.035.

[3] Creamer, J., Bristow, D.A. & Landers, R.G., Selection of limited and constrained compensation tables for five axis machine tools. *Int. J. Adv. Manuf. Technol.*, **92**, pp. 1315–1327, 2017.

[4] Kvrgic, V.M., Ribic, A.I., Dimic, Z., Zivanovic, S.T., Dodevska, Z.A., Equivalent geometric errors of rotary axes and novel algorithm for geometric errors compensation in a nonorthogonal five axis machine tool. *CIRP J. Manuf. Sci. Technol.*, **37**, pp. 477–488, 2022.

[5] Chen, P., Wang, S., Li, B. & Li, F., A novel geometric error compensation method for improving machining accuracy of spiral bevel gear based on inverse kinematic model. *Int. J. Adv. Manuf. Technol.*, **127**(9), pp. 4339–4355, 2023.

[6] Fu, W., Wang, B., Li, X., Liu, L. & Wang, Y., Ascent trajectory optimization for hypersonic vehicle based on improved chicken swarm optimization. *IEEE Access*, **7**, 151836–151850, 2019.

[7] Zhang, S., He, C., Liu, X., Xu, J. & Sun, Y., Kinematic chain optimization design based on deformation sensitivity analysis of a five axis machine tool. *Int. J. Precis. Eng. Manuf.*, **21**(12), pp. 2375–2389, 2020. https://doi.org/10.1007/s12541-020-00421-w.

[8] Maletić, J. & Živanović, S., Verification of inverse kinematic equations for a five axis machine tool with a spindle tilting configuration. *Adv. Technol. Mater.*, **47**(1), pp. 33–38, 2022. https://doi.org/10.24867/atm-2022-1-006.

[9] Maletić, J. & Živanovi, S., A post-processor for the five-axis machine MultiProDesk based on inverse kinematic transformation. *16th Int. Conf. on Accomplishments Mech. Ind. Eng.*, pp. 28–34, 2023.

[10] Ma, H.Y., Yuan, C.M., Shen, L.Y. & Gao, X.S., Optimal feedrate planning on a five axis parametric tool path with global geometric and kinematic constraints. *J. Comput. Des. Eng.*, **9**(6), pp. 2355–2374, 2022. https://doi.org/10.1093/jcde/qwac116.

[11] Ullah, A., Chan, T.-C. & Chang, S.-L., Enhancing five axis machine tool performance through esg-based design optimization. *Int. J. Precis. Eng. Manuf. Technol.*, pp. 1–18, 2024.

[12] Yan, Y., Zhang, L. & Gao, J., Tool path planning for flank milling of non-developable ruled surface based on immune particle swarm optimization algorithm. *Int. J. Adv. Manuf. Technol.*, **115**(4), pp. 1063–1074, 2021.

[13] Zhang, H., Xiang, S., Zhou, T. & Yang, J., Joint sensitivity analysis method of impeller aerodynamic performance to key geometric errors of five axis machine tool. *Eng. Sci. Technol. an Int. J.*, **57**, 101814, 2024. https://doi.org/10.1016/j.jestch.2024.101814.

[14] Gao, S., Zhou, H., Hu, P., Chen, J., Yang, J. & Li, N., A general framework of workpiece setup optimization for the five axis machining. *Int. J. Mach. Tools Manuf.*, **149**, 103508, 2020.

[15] Lu, Z., Huo, G. & Jiang, X., A novel method to minimize the five axis CNC machining error around singular points based on closed-loop inverse kinematics. *Int. J. Adv. Manuf. Technol.*, **128**(5–6), pp. 2237–2249, 2023.

[16] Dening, S., Chao, Z., Jianwei, M., Jinghua, L. & Yuguang, Z., An adaptive singularity avoidance method for five axis toolpath generation with invariant cutter-contact points and bounded tool orientation variation based on the reduction of Jacobian condition number. *J. Manuf. Process.*, **101**, pp. 1521–1537, 2023.

[17] Wei, J., Jiang, L. & Liang, B., Geometric error compensation method of five axis CNC machine tool based on improved genetic algorithm. *Journal of Physics: Conference Series*, 12029, 2024. https://doi.org/10.1088/1742-6596/2825/1/012029.

[18] Murray, R.M., Li, Z. & Sastry, S.S., *A Mathematical Introduction to Robotic Manipulation*, CRC Press, 2017.
[19] Xiang, S. & Altintas, Y., Modeling and compensation of volumetric errors for five axis machine tools. *International Journal of Machine Tools and Manufacture*, **101**, pp. 65–78, 2016.
[20] Goldberg, D.E. & Richardson, J., Genetic algorithms with sharing for multimodal function optimization. *Genetic Algorithms and Their Applications: Proceedings of the Second International Conference on Genetic Algorithms*, **4149**, pp. 414–425, 1987.
[21] Holland, J.H., Genetic algorithms. *Scientific American*, **267**(1), pp. 66–73, 1992.
[22] Ahmad, M., Kumar, N. & Kumari, R., A hybrid genetic algorithm approach to solve inverse kinematics of a mechanical manipulator. *International Journal of Scientific and Technology Research*, **8**(9), pp. 1777–1782, 2019.
[23] Spaney, P. et al., A model-driven digital twin for manufacturing process adaptation. *Proceedings of the 2023 ACM/IEEE International Conference on Model Driven Engineering Languages and Systems Companion (MODELS-C)*, pp. 465–469, 2023.
[24] Zenhari, S., Möhring, H.C. & Torbati, A.V., Comprehensive analysis of kinematic models based on the DH method and screw theory for a five axis machine tool. *Procedia CIRP*, **130**, pp. 1745–1751, 2024.

SECTION 7
SHAPE AND TOPOLOGY OPTIMISATION

TOPOLOGY OPTIMIZATION OF STRUCTURES WITH STRESS, BUCKLING AND DYNAMIC BEHAVIOUR CONSTRAINTS

MARTÍN REY, IVAN COUCEIRO, JOSÉ PARÍS, LUIS RAMÍREZ & FERMÍN NAVARRINA
Group of Numerical Methods in Engineering, Center for Technological Innovation in Construction and Civil Engineering, Civil Engineering School, Universidade da Coruña, Spain

ABSTRACT

We can define topological optimization of structures as the branch of computational mechanics that seeks to obtain optimal material layouts in a given domain under specific boundary conditions. This discipline, pioneered by Bendsøe and Kikuchi in the late 1980s, is becoming increasingly relevant due to its direct application in various sections of applied physics such as civil, mechanical and aeronautical engineering. In recent years, the advancements in computational capacity have enabled a significant leap in the topological optimization of structures. This work aims, first, to systematize and compile the current knowledge surrounding the behaviour of structural models with intrinsic relative densities used in the SIMP method, in order to design solutions to the various physical and numerical challenges arising from the use of Galerkin interpolation methods (FEM, IGA, etc.) for solving elasticity equations with relative density. Specifically, we will attempt to address the problems arising from topological optimization processes in which dynamic structural analysis or second-order theory plays a fundamental role. Ultimately, the main objective of this work is to design topological optimization formulations for structures of isotropic materials using the SIMP method integrating advanced constraints and combining them in a way that accounts for an increasing number of physical phenomena, thereby reducing the structure's weight to optimal level. The reduction in structural weight introduces the potential for instabilities, dynamic effects and nonlinearities, which will be addressed throughout the optimization process using well-known structural analysis techniques such as the calculation of natural vibration frequencies, linear buckling or full geometric nonlinearity.

Keywords: structures, optimization, buckling, frequencies, stress.

1 INTRODUCTION

Topological optimization of structures is a discipline that began to take its first steps in the 1980s, with the development of computers and the evolution of the finite element method [1]. This discipline can be understood as the intersection of three distinct scientific branches: on one hand, optimization, a field that aims to minimize or maximize an objective function in a given problem, subject (or not) to certain constraints; on the other hand, topology, a branch of mathematics that studies the properties of continuous geometric bodies; and finally, structural analysis, a branch of engineering that seeks to determine the behaviour of a specific structure using physical principles.

Based on this foundation, this discipline can be defined as the combination of the three aforementioned branches, as it seeks to maximize or minimize a property of a structure without imposing a discontinuity or abrupt change in it, that is, while maintaining a global behaviour. The usefulness of topological optimization of structures has been widely demonstrated in all fields of engineering, from industrial engineering – where the automated execution of industrial components makes optimization play a fundamental role – to civil engineering, where reducing material costs is one of the primary objectives.

As of today, advancements in structural topology optimization have made it possible to integrate greater control over the dynamic and energy stability of structures into these processes. Significant progress has been made toward achieving a comprehensive

WIT Transactions on The Built Environment, Vol 216, © 2025 WIT Press
www.witpress.com, ISSN 1743-3509 (on-line)
doi:10.2495/HPSM250121

optimization process that considers both stress and stability factors. However, there is still a need to develop a global optimization framework for steel structures that brings us closer to bridging the gap between mathematical optimization and the design of realistic structures that comply with established regulations.

2 STRUCTURAL RESPONSES

2.1 Introduction to structural responses

When we calculate and design structures, simplifications can be made by assuming that the state in the deformed position does not change or hardly changes compared to the initial geometric arrangement. The formulation of this hypothesis allows us to simplify structural calculations while obtaining highly accurate results for specific cases. However, when dealing with slender structures, neglecting nonlinear behaviour can lead to significant errors in the final results. Therefore, we must calculate the structure while considering that the deformed position will differ significantly from the initial position.

2.2 Geometrically non-linear formulation

First, it is important to highlight that the structural analysis method used in this work is linear. However, the explanation of the nonlinear formulation is necessary because the implementation of linear instability, which will be addressed in later sections, as well as the linear formulation itself and dynamic analysis, are based on the development of the updated Lagrangian formulation, which we will now present. So, we begin with the incremental expression of Principle of Virtual Work:

$$\int_{V_0} \delta\boldsymbol{\gamma}^T \boldsymbol{\tau}^{t+\Delta t} dV = \int_{S_0} \delta\boldsymbol{u}^T \boldsymbol{t_0}\, dS + \int_{V_0} \rho_0 \delta\boldsymbol{u}^T \boldsymbol{g}\, dV \tag{1}$$

where the right-hand side corresponds to the terms of the virtual external work, with $\boldsymbol{\gamma}$ as the Euler–Lagrange strain tensor, $\boldsymbol{\tau}$ as the second Piola–Kirchoff stress tensor, $\boldsymbol{t}$ being the nominal forces on the boundary of the continuous medium, $\boldsymbol{g}$ the gravitational forces, and $\boldsymbol{u}$ the displacements.

Since we will not discuss plasticity in this work, we can establish a linear relationship between the stress increment of the second Piola–Kirchhoff tensor and the Green–Lagrange strain tensor. In summary, we can introduce the material's constitutive equation, which for linear elasticity will be the well-known Hooke's Law:

$$\Delta\boldsymbol{\tau} = \boldsymbol{D}\Delta\boldsymbol{\gamma} \tag{2}$$

where $\boldsymbol{D}$ is the constitutive tensor.

Since we are working in a nonlinear formulation, the increment of the Green–Lagrange strain tensor will have a linear part and a nonlinear part. Discarding the nonlinear part of the tensor and rearranging the terms, we obtain the following:

$$\int_{V_0} \delta\Delta\boldsymbol{e}^T \boldsymbol{D}\, \Delta\boldsymbol{e}\, dV + \int_{V_0} \delta\boldsymbol{\eta}^T \boldsymbol{\tau}^t\, dV = \int_{S_0} \delta\boldsymbol{u}^T \boldsymbol{t_0}\, dS + \int_{V_0} \rho_0 \delta\boldsymbol{u}^T \boldsymbol{g}\, dV - \int_{V_0} \delta\boldsymbol{e}^T \boldsymbol{\tau}^t\, dV \tag{3}$$

where $\boldsymbol{e}$ and $\boldsymbol{\eta}$ are the linear and non-linear parts of the Green–Lagrange strain tensor respectively. Having obtained the present equation, it is important to note that we will use the finite element method for the discretization of the problem.

Thus, we will obtain an implicit iterative scheme where the resolution of each step can be carried out through a Newton–Raphson scheme, displacement control, or the most suitable method for the problem at hand. After applying the FEM, we obtain the following:

$$(\boldsymbol{K_L} + \boldsymbol{K_G})\,\Delta\boldsymbol{a} = \left(\boldsymbol{f}_{\mathbf{ext}}^{t+\Delta t} + \boldsymbol{f}_{\mathbf{int}}^{t}\right) \tag{4}$$

$$\boldsymbol{f}_{\mathbf{ext}}^{t+\Delta t} = \int_{S_0} \boldsymbol{H}^{\mathrm{T}}\boldsymbol{t_0}\,\mathrm{d}S + \int_{V_0} \rho_0 \boldsymbol{H}^{\mathrm{T}}\boldsymbol{g}\,dV\,;\ \boldsymbol{f}_{\mathbf{int}}^{t} = \int_{V_o} \boldsymbol{B}_L^T\boldsymbol{\tau}^t\,dV$$

$$\boldsymbol{K_L} = \int_{V_0} \boldsymbol{B}_L^T\boldsymbol{D}\boldsymbol{B}_L dV\ ;\boldsymbol{K_G} = \int_{V_0}^{t} \boldsymbol{B}_{NL}^T\boldsymbol{T}\boldsymbol{B}_{NL} dV$$

where $\boldsymbol{a}$ are the nodal displacements of the mesh, $\boldsymbol{K_L}$ the linear stiffness matrix, $\boldsymbol{K_{NL}}$ the geometric stiffness matrix (the sum of both is called the tangent matrix), $\boldsymbol{f}_{\mathbf{ext}}^{t+\Delta t}$ the external forces of the structure at the final of the load step, $\boldsymbol{f}_{\mathbf{int}}^{t}$ the internal forces of the structure and $\boldsymbol{H}, \boldsymbol{B}_L, \boldsymbol{B}_{NL}$ are the matrices that contain, respectively, the shape functions of the vector space and their derivatives. For a much more exhaustive explanation of this and the following formulations, it is recommended to consult [2].

Following this scheme, what we are doing is minimizing the residual of the structure. In equilibrium for predetermined loads, the following occurs:

$$\boldsymbol{f}_{\mathbf{ext}}^{end} = \boldsymbol{f}_{\mathbf{int}}^{end} \tag{5}$$

And taking this into account, if (for example) we solve the equation using the Newton–Raphson method, the tangent stiffness matrix is nothing more than the derivative of the residual we want to minimize and, therefore, the derivative of the internal force with respect to the displacements, since in most cases, the derivative of the external forces does not depend on the displacements.

Thus, if we wanted to introduce a variable such as relative density into the formulation, the proper approach would first be to implement it in the computation of the internal forces due to the volume integral performed to obtain the forces, which depends on the density of each element. Thus, developing the by-element formulation:

$$\boldsymbol{f}_{\boldsymbol{int,e}}(\rho_e, \mathbf{u_e}) = \int \boldsymbol{\rho}_{\mathbf{e}}\, \boldsymbol{B}_{L,e}^T\boldsymbol{\tau}_e^t dV \rightarrow \mathbf{K}_{\mathbf{m,e}} = \frac{\partial \boldsymbol{f}_{\boldsymbol{int,e}}}{\partial \boldsymbol{u}} = \rho_e \boldsymbol{K_m} \tag{6}$$

with $\boldsymbol{K_m}$ as the tangent stiffness matrix. Thus, considering the previous development and assembling the elemental matrices, we obtain the following formulation of the matrices for meshes where the elements have relative density.

$$\boldsymbol{K_m}\,\Delta\boldsymbol{a} = \left(\boldsymbol{f}_{\mathbf{ext}}^{t+\Delta t} + \boldsymbol{f}_{\mathbf{int}}^{t}\right) \tag{7}$$

$$\boldsymbol{f}_{\mathbf{ext}}^{t+\Delta t} = \int_{S_0} \boldsymbol{H}^{\mathrm{T}}\boldsymbol{t_0}\,\mathrm{d}S + \int_{V_0} \rho_0 \boldsymbol{H}^{\mathrm{T}}\boldsymbol{g}\,\boldsymbol{\delta}_e\boldsymbol{\rho}_e\,dV\,;\ \boldsymbol{f}_{\mathbf{int}}^{t} = \int_{V_o} \boldsymbol{B}_L^T\boldsymbol{\tau}^t\ \boldsymbol{\delta}_e\boldsymbol{\rho}_e\,dV$$

$$\boldsymbol{K_L} = \int_{V_0} \boldsymbol{B}_L^T\boldsymbol{D}\boldsymbol{B}_L\ \boldsymbol{\delta}_e\boldsymbol{\rho}_e\,dV\,;\boldsymbol{K_G} = \int_{V_0}^{t} \boldsymbol{B}_{NL}^T\boldsymbol{T}\boldsymbol{B}_{NL}\ \boldsymbol{\delta}_e\boldsymbol{\rho}_e\,dV\,;\boldsymbol{K_m} = \boldsymbol{K_L} + \boldsymbol{K_{NL}}$$

2.3 Linear formulation

The finite element method in the linear regime will be used to calculate the structures within this project, so we will define it up to the calculation of the system of linear equations to be implemented. We will start from all the previously developed formulation, which will also be necessary in later sections. For the formulation of the finite element method in the linear regime, we can make a series of simplifications based on the statement of the principle of virtual work. The simplifications are as follows:

- Displacements are small.
- The displacement gradient is also small.
- The material is elastic and linear.

Under these simplifications, the formulation remain like it follows:

$$\boldsymbol{K}\,\boldsymbol{a} = \boldsymbol{f} \tag{8}$$

$$\boldsymbol{f} = \int_{S_0} \boldsymbol{H}^{\mathrm{T}} \boldsymbol{t_0}\, \mathrm{d}S + \int_{V_0} \rho_0 \boldsymbol{H}^{\mathrm{T}} \boldsymbol{g}\, \boldsymbol{\delta_e \rho_e}\, dV$$

$$\boldsymbol{K_L} = \int_{V_0} \boldsymbol{B}^T \boldsymbol{D} \boldsymbol{B}\, \boldsymbol{\delta_e \rho_e}\, dV$$

where $\boldsymbol{B}$ is the matrix that contain the derivatives of the shape functions. The difference between this matrix and $\boldsymbol{B_L}$ is that $\boldsymbol{B}$ does not depend on the strain gradient.

2.4 Linear buckling analysis

The method of linear instability is based on the existence of the equilibrium bifurcation point. That is, along the force–displacement curve of the structure, there is a point at which the trajectory splits into a basic equilibrium branch and an instability branch. This method is used to obtain the bifurcation points of instability. By simplifying the non-linear formulation, we can obtain the bifurcation points by solving an eigenvalue problem like:

$$(\boldsymbol{K} + \lambda_i \boldsymbol{K_{NL}})\boldsymbol{\phi}_i = \boldsymbol{0} \tag{9}$$

where $\boldsymbol{K}$ and $\boldsymbol{K_{NL}}$ are the same matrix as in the previous sections.

In the case of applying the linear instability method to models with relative densities, we must be careful with the appearance of spurious buckling modes. These buckling modes appear due to an artificial increase in stress in areas with very low relative densities. This formulation does not pose any numerical problems when the constraints to be considered in the optimization problem are not intrinsic to the material's density and its effect on the different structural responses. However, in problems related to eigenvalues and eigenvectors, where the material's own density conditions the final calculations, this effect can lead to numerical instabilities [3].

2.5 Dynamics and natural frequencies of vibrations

The natural frequencies of a structure are those vibration frequencies at which a structure continues to vibrate naturally, that is, the ones that indicate the vibration modes of a structure. They are closely related to the number of degrees of freedom of the structure itself, as the number of natural frequencies of a structure is equal to its number of degrees of freedom. Each natural frequency is associated with a vibration mode, in such a way that the structure vibrates differently according to each frequency.

We can define the matrix equation that governs the motion problem of a body with multiple degrees of freedom (without considering damping forces) [4]:

$$\boldsymbol{M}\ddot{\boldsymbol{u}} + \boldsymbol{K}\boldsymbol{u} = \boldsymbol{f}(t) \tag{10}$$

If we assume that the movements of the structure will follow the following scheme $[u(t) = \boldsymbol{\phi}e^{i\omega t}; \ddot{u} = -\omega^2\, \boldsymbol{\phi}e^{i\omega t}]$, we obtain the eigenvalue formulation that allows us to solve the natural vibration modes of the structure, as we can see in the following development:

$$(\boldsymbol{K} + \omega_i^2\, \boldsymbol{M})\boldsymbol{\phi}_i = \boldsymbol{0} \tag{11}$$

where $\boldsymbol{M}$ is the mass matrix of the structure which, considering the implementation of relative densities, can be calculated as:

$$\boldsymbol{M} = \int_{V_0} \boldsymbol{H}^T \rho_0 \boldsymbol{H}\ \boldsymbol{\delta_e \rho_e} dV \tag{12}$$

3 OPTIMIZATION PROBLEM, OBJECTIVE FUNCTION AND CONSTRAINTS

As we mentioned earlier, the optimization problem we are going to solve focuses on obtaining the material distribution in a specific domain under a set of imposed loads, minimizing the weight of the structure, and potentially being subject to constraints on stress, buckling, or frequencies. The general mathematical formulation that encompasses all cases of this problem is the following:

$$\begin{aligned}
&\text{find:} && \boldsymbol{\rho} = \{\rho_i\} \\
&\text{that minimise:} && F(\boldsymbol{\rho}) \\
&\text{verifying:} && g_j(\boldsymbol{u}, \boldsymbol{\rho}) \ \leq 0 \\
& && h_1(\boldsymbol{u}, \boldsymbol{\rho}) \ \leq 0 \\
& && h_2(\boldsymbol{u}, \boldsymbol{\rho}) \ \leq 0 \\
&\text{subject to:} && \boldsymbol{K}\,\boldsymbol{u} = \boldsymbol{f} \\
& i = 1, \dots, N_e \ ; && j = 1, \dots, N_{\text{rest}}
\end{aligned} \tag{13}$$

where $\boldsymbol{\rho}$ is the vector of design variables, $F(\boldsymbol{\rho})$ is the objective function, $g_j(\boldsymbol{u}, \boldsymbol{\rho})$ are the local constraints (such as tensions, in our case), and $h_k(\boldsymbol{u}, \boldsymbol{\rho})$) are the global constraints, such as linear buckling constraints or eigenfrequencies constraints. Some studies have partially used this formulation (with different objective functions and modifications in the constraints), such as [5]–[7].

3.1 Objective function

As an objective function, as indicated in previous sections, we will use weight. A weight minimization formulation that does not penalize intermediate densities may lead to the appearance of areas in the domain with a high predominance of these densities. To correct this effect, a modified objective function is proposed, in which the cost is penalized. Thus, the use of the objective function proposed by Chopra [4] is suggested, which penalizes intermediate densities through a factor that does not affect the final behaviour of the structure.

$$F = \sum_{e=1} \int_{V_e} (\rho_e)^{1/p} \, dV \tag{14}$$

where ρ_e is the relative density of a certain element and p is the penalization factor.

3.2 Stress constraints

The stress constraints at the centroid of each element must follow specific criteria during their definition. In the current structural calculation model, there are six stress values calculated at each Gaussian integration point: $\sigma_x, \sigma_y, \sigma_z, \tau_{xy}, \tau_{xz}, \tau_{yz}$. These stresses form the stress tensor of the structure. The constraint will be implemented through an equivalent stress criterion. Since the material in the proposed formulation is steel, the stress used will be the Von Mises equivalent stress:

$$\sigma_{vm} = \sqrt{\frac{1}{2}\left[\left(\sigma_x - \sigma_y\right)^2 + \left(\sigma_y - \sigma_z\right)^2 + (\sigma_z - \sigma_x)^2 + 6\left(\tau_{xy}^2 + \tau_{yz}^2 + \tau_{zx}^2\right)\right]} \tag{15}$$

In this way, we can define the stress constraints as:

$$g_e(\boldsymbol{\rho}) = \sigma_{vm,e}\left(\boldsymbol{\sigma}^h(\boldsymbol{u}_e^o, \rho)\right) - \sigma_{vm,\max} \leq 0 \tag{16}$$

where $g_e(\boldsymbol{\rho})$ is the constraint, $\sigma_{vm,e}$ is the Von Mises equivalent stress at the centroid of the element and $\sigma_{vm,\max}$ is the limit of the constraint.

3.3 Buckling load factor constraint

The linearized buckling constraint is a global constraint, meaning that it is obtained collectively for the entire structure through the model presented in this work. Thus, through the previously presented eigenvalue problem, it is possible to determine the linearized buckling factor λ, which is used in the constraint, resulting in the following form:

$$h_{\text{buck}}(\rho) = \lambda_{\text{lim}} - \lambda(\boldsymbol{u}, \boldsymbol{\rho}) \leq 0 \tag{17}$$

where λ_{lim} is the limit imposed on the first factor of linearized buckling.

3.4 Natural frequencies constraints

Similar to the case of linearized buckling, this constraint is also global in nature and is calculated through the model proposed and validated in this work. While, in the case of the buckling factor, there was an interest in restricting the value from below, this is not the case here, since the limit of the first natural frequency of vibration can depend on multiple factors, ranging from the structural objective to the dynamic loads it must withstand throughout its service life. In some cases, these limits on natural frequencies may be imposed by the relevant regulations in each structural domain. In this way, we obtain two natural frequency constraints that can be used separately or together, depending on the problem being posed.

$$h_{\text{inf,frec}}(\rho) = \lambda_{\text{lim}}^{\text{inf}} - \lambda(\boldsymbol{u}, \boldsymbol{\rho}) \leq 0 \tag{18}$$

$$h_{\text{sup,frec}}(\rho) = \lambda(\boldsymbol{u}, \boldsymbol{\rho}) - \lambda_{\text{lim}}^{\text{sup}} \leq 0 \tag{19}$$

where $h_{\text{inf,frec}}$ is the lower limit imposed on the first natural frequency, and $h_{\text{sup,frec}}$is the upper limit.

4 SENSITIVITY ANALYSIS

The optimization process presented in previous sections, in most cases and depending on the resolution algorithm used, requires the computation of the derivatives of the constraints and the objective function to be implemented. Thus, its proper functioning depends on the accuracy of these derivative calculations and the quality of the data provided in this regard. This process of computing derivatives with respect to the design variables in an optimization problem is known as sensitivity analysis and is one of the key components of the problem. It consumes a very high percentage of the computation time and, in turn, significantly influences our ability to reach the global optimum of the problem. In our case, we will present the process for obtaining the first-order derivatives of the functions involved in the problem so that they can be incorporated into the formulation.

4.1 Sensitivity of the objective function

Recalling the previously defined objective function, it is straightforward to obtain that its sensitivity with respect to the relative density of any element is:

$$\frac{dF}{d\rho_e} = \int_{V_e} \frac{1}{p} \rho_e^{\left(\frac{1}{p}-1\right)} dV \tag{20}$$

4.2 Stress sensitivity

Since we have defined the stress to be used in the optimization problem as the von Mises comparison stress, we need its derivative. We will express it using the chain rule as follows:

$$\frac{dt}{da} = \left.\frac{d\sigma_{vm}}{d\sigma}\right|_{\sigma(a)} \frac{d\sigma}{da}\frac{da}{d\rho_i} + \frac{dt(\rho_i)}{d\rho_i} \tag{21}$$

where $\boldsymbol{t}$ is the stress constraint. The last term must be zero in our case because there is not direct dependency between the constraint and the relative density. First, we can define the sensitivity of the Von Mises stress with respect to the stress tensor as:

$$\frac{d\sigma_{vm}}{d\sigma} = \begin{bmatrix} \frac{\partial\sigma_{vm}}{\partial\sigma_x} & \frac{\partial\sigma_{vm}}{\partial\sigma_y} & \frac{\partial\sigma_{vm}}{\partial\tau_{xy}} \end{bmatrix} \tag{22}$$

where:

$$\frac{\partial\sigma_{vm}}{\partial\sigma_x} = \frac{2\sigma_x - \sigma_y}{2\sigma_{vm}}; \frac{\partial\sigma_{vm}}{\partial\sigma_y} = \frac{2\sigma_y - \sigma_x}{2\sigma_{vm}}; \frac{\partial\sigma_{vm}}{\partial\tau_{xy}} = \frac{6\tau_{xy}}{2\sigma_{vm}} \tag{23}$$

In the formulation of the finite element method, the stress tensor of a continuous medium is calculated as:

$$\boldsymbol{\sigma}(\boldsymbol{r}) = \boldsymbol{DBu}(\boldsymbol{r}) \tag{24}$$

where $\boldsymbol{u}(\boldsymbol{r})$ are the displacements at point $\boldsymbol{r}$. Applying the discretization, we can also define the tensor as:

$$\boldsymbol{\sigma}(\boldsymbol{r}) = \sum_{i=1}^{n} \boldsymbol{DB}\,\boldsymbol{H}_i(\boldsymbol{r})\boldsymbol{a}_i \tag{25}$$

Thus, we can formulate the derivatives of the stress tensor with respect to the nodal displacements of the structure as:

$$\frac{d\boldsymbol{\sigma}}{da_i} = \boldsymbol{DBH}_i(\boldsymbol{r})\boldsymbol{\delta}_i \tag{26}$$

where $\boldsymbol{D}$ is the constitutive matrix, $\boldsymbol{B}$ is the matrix of shape function derivatives, and $\boldsymbol{H}$ is the shape function matrix.

In order to obtain the last term, we can start from eqn (8). Differentiating this expression, we obtain:

$$\boldsymbol{K}\frac{d\boldsymbol{a}}{d\rho_i} = \frac{d\boldsymbol{f}}{d\rho_i} - \frac{d\boldsymbol{K}}{d\rho_i}\boldsymbol{a} \tag{27}$$

from which, by solving, we can obtain the desired term:

$$\frac{d\boldsymbol{a}}{d\rho_i} = \boldsymbol{K}^{-1}\left[\frac{d\boldsymbol{f}}{d\rho_i} - \frac{d\boldsymbol{K}}{d\rho_i}\boldsymbol{a}\right] \tag{28}$$

The full development of this formulation can be found in París et al. [8].

4.3 Sensitivity of the natural vibrations

We will start from the development defined in the previous section to obtain the sensitivity of the first natural vibration frequency of the structure. The sensitivity formulation of the eigenvalue problems has been obtained following Hernández [9]. To obtain the natural frequencies, it was necessary to solve the eigenvalue problem defined previously:

$$(\boldsymbol{K} - \omega^2\boldsymbol{M})\boldsymbol{\phi} = 0 \tag{29}$$

However, our intention was not to obtain all the natural vibration frequencies but rather the smallest one in absolute value. Therefore, we reformulated the problem by $\nu_i = \frac{1}{{\omega_i}^2}$, leading to:

$$(\boldsymbol{M} - \nu_i\boldsymbol{K})\boldsymbol{\phi}_i = 0 \tag{30}$$

For simplicity, we are going to omit the subscript i from the natural frequency and instead use it in reference to the design variables. By differentiating the previous expression, we obtain the following development:

$$\boldsymbol{\phi}^t\left[\frac{\partial\boldsymbol{M}}{\partial\rho_i} - \frac{\partial\nu}{\partial\rho_i}\boldsymbol{K} - \nu\frac{\partial\boldsymbol{K}}{\partial\rho_i}\right]\boldsymbol{\phi} = 0 \tag{31}$$

Thus, we obtain the expression that gives us the sensitivity of the natural vibration frequency:

$$\frac{\partial\nu}{\partial\rho_i} = \boldsymbol{\phi}^t\left[\frac{\partial\boldsymbol{M}}{\partial\rho_i} - \nu\frac{\partial\boldsymbol{K}}{\partial\rho_i}\right]\boldsymbol{\phi} \tag{32}$$

However, we cannot forget about the initial change made, nor that we are looking for the natural dimensional frequency, in Hz, so we need to make the following changes:

$$f = \frac{1}{2\pi}\sqrt{\frac{1}{\nu}} \quad \rightarrow \quad \frac{\partial f}{\partial \rho_i} = -\frac{1}{4\pi}\nu^{-\frac{3}{2}}\frac{\partial \nu}{\partial \rho_i} \tag{33}$$

4.4 Sensitivity of the buckling load factor

The formulation of buckling load factor is very similar in its development to that of frequency calculation, with the exception of the subsequent treatment of the derivative of the matrices. Thus, we start from the development defined in previous sections:

$$[\boldsymbol{K} + \lambda_i \boldsymbol{K_G}]\boldsymbol{\phi}_i = 0 \tag{34}$$

However, just as in the case of frequencies, we need to obtain the minimum factor in absolute value, so the actual development will be through the inverse Mises method by defining the variable $\nu_i = \frac{1}{\lambda}$

$$[\boldsymbol{K}_G + \nu_i \boldsymbol{K}]\boldsymbol{\phi}_i = 0 \tag{35}$$

In this case, we will also omit the subscript referring to the buckling mode and will use it in reference to the design variables. Differentiating, we obtain:

$$\boldsymbol{\phi}^t \left[\frac{\partial \boldsymbol{K_G}}{\partial \rho_i} + \frac{\partial \nu}{\partial \rho_i}\boldsymbol{K} + \nu\frac{\partial \boldsymbol{K}}{\partial \rho_i}\right]\boldsymbol{\phi} = 0 \tag{36}$$

Thus, we obtain the expression that gives us the sensitivity of the first inverse linearized buckling factor:

$$\frac{\partial \nu}{\partial \rho_i} = -\boldsymbol{\phi}^t \left[\frac{\partial \boldsymbol{K_G}}{\partial \rho_i} + \nu\frac{\partial \boldsymbol{K}}{\partial \rho_i}\right]\boldsymbol{\phi} \tag{37}$$

However, we must not forget that we are seeking the derivative of the linearized buckling factor with the eigenvalue problem, therefore

$$\frac{\partial \lambda}{\partial \rho_i} = \frac{\partial \lambda}{\partial \nu}\frac{\partial \nu}{\partial \rho_i} = \frac{1}{\nu^2}\boldsymbol{\phi}^t \left[\frac{\partial \boldsymbol{K_G}}{\partial \rho_i} + \nu\frac{\partial \boldsymbol{K}}{\partial \rho_i}\right]\boldsymbol{\phi} \tag{38}$$

5 BENCHMARK CASES AND IMPLEMENTATION

Once the formulation has been presented, we will develop a series of benchmark cases to observe compliance with the constraints and the weight reduction of two structures. The ARPACK library [10] has been used to obtain the eigenvalues associated with each problem. For the process of obtaining the structural responses, proprietary software written in FORTRAN has been used, as well as for solving the optimization process, where sequential linear programming has been applied. To eliminate stress singularities and spurious buckling modes, the stress relaxation formulation [8], [11] has been implemented.

We will test two benchmark cases, both related to the reduction of the steel structure's weight but using a compression column to constrain the buckling factor and an MBB beam to impose frequency constraints. We have run the benchmark cases with frequency constraints of 140/150 Hz and buckling factor constraints of 40/50. In Figs 1–4, the evolution of the structure's weight and the relevant constraints can be observed. The computation time is significantly higher (two orders of magnitude greater) in the case of the compression column, as the sensitivity of the buckling factor is computationally much more expensive than the sensitivity of the frequencies.

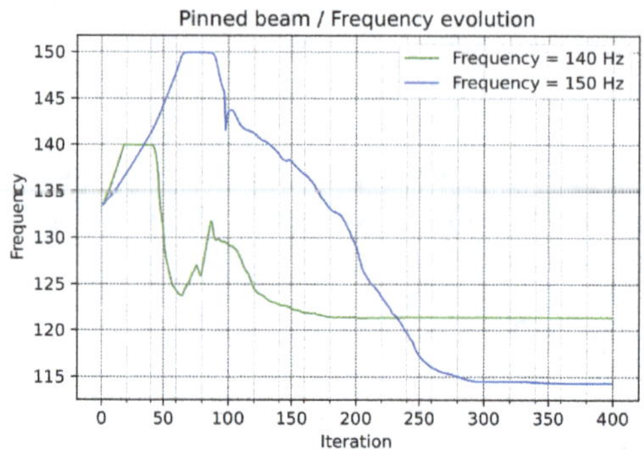

Figure 1: Beam frequency evolution.

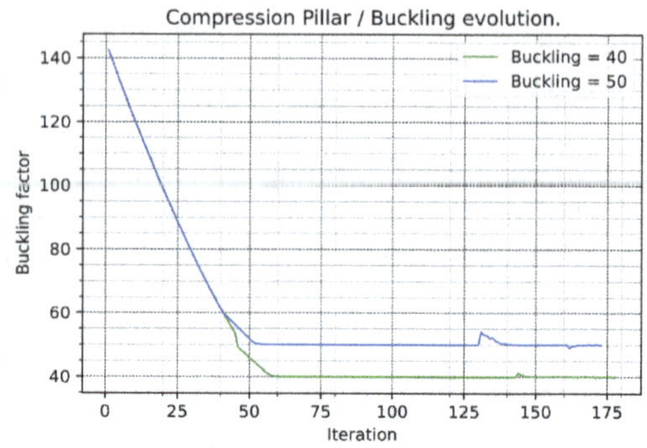

Figure 2: Pillar buckling evolution.

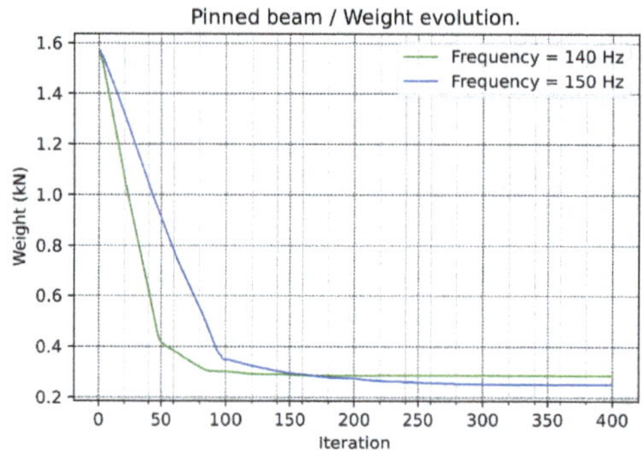

Figure 3: Beam weight evolution.

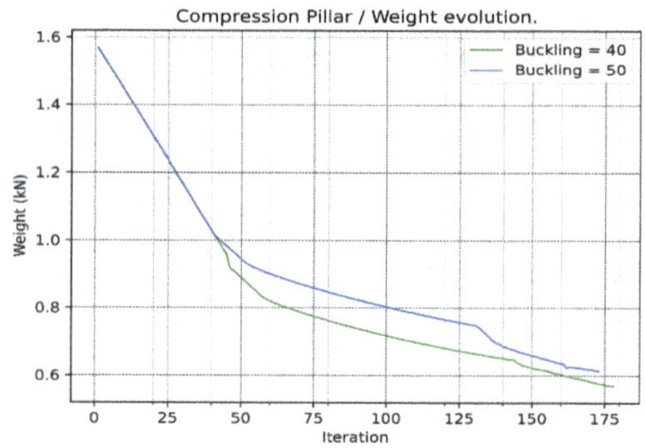

Figure 4: Pillar weight evolution.

6 CONCLUSIONS

This work presents the formulation required to carry out a topology optimization process using the SIMP method, starting from the linear and nonlinear structural response of the structure to the sensitivity of all functions involved in the problem. A general optimization formulation has been obtained, combining local and global constraints related to both stress and energy and dynamic instability. The work has focused solely on the physical problem, not on the numerical issues arising from the implementation of this type of formulation. The dynamic and buckling analysis is obtained through eigenvalue problems, which have also been studied to analytically determine their sensitivity.

Future research directions should consider the combined behaviour of buckling and natural vibration frequencies, as well as their mutual couplings, alongside improvements in computational efficiency. Additionally, a deeper study of existing works on numerical instabilities and spurious buckling modes is necessary. Furthermore, the integration of material nonlinearities should be addressed to ensure the proper performance of the structure, especially if geometric nonlinearity is to be incorporated during the optimization process.

ACKNOWLEDGEMENTS

This research has been partially funded by strategic projects oriented towards the ecological transition and the digital transition of the Ministry of Science and Innovation: GREEN-HUGS (Grant #Ted2021-129991b-C31 and Grant #Ted2021-129991b-C32) and NEPTUNE (Grant #Ted2021-129805b-I00), by The Ministry of Science and Innovation: ATHENEA (Grant #Pid2021-125447ob-100) and TOPACIUS (Grant #Pid2023-146793ob-I00), by Department of Education and University Planning of The Xunta De Galicia (Grant #Ed431c 2022/06), and by The Group of Numerical Methods in Engineering of The Universidade Da Coruña.

REFERENCES

[1] Bendsøe, M. & Sigmund, O., *Topology Optimization. Theory, Methods, and Applications*, 2nd ed., Springer, 2004.

[2] de Borst, R., Crisfield, M.A., Remmers, J.J.C. & Verhoosel, C.V., *Nonlinear Finite Element Analysis of Solids and Structures,* Wiley-Blackwell, 2012.

[3] Doan, Q.H. & Lee, D., Optimum topology design of multi-material structures with non-spurious buckling constraints. *Advances in Engineering Software*, **114**, pp. 110–120, 2017.

[4] Chopra, A.K., *Dynamic of Structures, Theory and Applications to Earthquake Engineering*, 3rd Ed., Prentice Hall, 2007.

[5] Ferrari, F. & Sigmund, O., Revisiting topology optimization with buckling constraints. *Structural and Multidisciplinary Optimization*, **59**(5), pp. 1401–1415, 2019.

[6] Li, Q., Wu, Q., Liu, J., He, J. & Liu, S., Topology optimization of vibrating structures with frequency band constraints. *Structural and Multidisciplinary Optimization*, **63**(3), pp. 1203–1218, 2020.

[7] Mitjana, F., Cafieri, S., Bugarin, F., Segonds, S., Castanie, F., & Duysinx, P., Topological gradient in structural optimization under stress and buckling constraints. *Applied Mathematics and Computation*, **409**, 126032, 2021.

[8] París, J. et al., Topology optimization of continuum structures with local and global stress constraints. *Struct. Multidisc. Optim.*, **39**, pp. 419–437, 2009.

[9] Hernández, S., Métodos de diseño óptimo de estructuras. Colegio Oficial de Ingenieros de Caminos, Canales y Puertos, 1990.

[10] Lehoucq, R.B., Sorensen, D.C. & Yang, C., *ARPACK: Solution of Large Scale Eigenvalue Problems by Implicitly Restarted Arnoldi Methods*, 1997.

[11] Xu, T., Huang, X., Lin, X. & Xie, Y.M., Topology optimization for maximizing buckling strength using a linear material model. *Computer Methods in Applied Mechanics and Engineering*, **417**, 2023.

NEW LENGTH-SCALE CONTROL APPROACHES FOR TOPOLOGY OPTIMISATION

ALAIN GARAIGORDOBIL, JOSÉ ANTONIO POSTIGO, RUBÉN ANSOLA & JAVIER CANALES
Department of Mechanical Engineering, School of Engineering of Bilbao, University of the Basque Country, Spain

ABSTRACT

Among the many advantages of the recent additive manufacturing technologies, certain limitations still hinder their widespread adoption in industry. Within the family of processes focusing on continuous fibre reinforcement, the inability to embed fibres in narrow regions is the quintessence of these issues. To address this issue and circumvent the need for a redesign-for-manufacturing phase, this work proposes a novel length-scale control algorithm that produces blur print designs. The present work develops two alternative and integrable minimum length-scale processes: a local perimeter controlling method, and a skeleton-based method. Both approaches leverage commonly used filtering strategies, ensuring reduced computational times while effectively deriving the structure's skeleton.
Keywords: topology optimisation, additive manufacturing, continuous fibre, length-scale control.

1 INTRODUCTION

This article presents a new methodology for controlling the minimum size of structural elements in topology optimisation. The proposed approach utilises common filtering techniques employed in topology optimisation, thereby eliminating the need for additional finite element analyses.

One of the first contributions to scale control was made by Guest et al. [1] who proposed a method for enforcing a minimum length scale by projecting the nodal design variable onto the element field within a filter radius r_{min}. This approach ensures control over the minimum size of solid elements. On the other hand, Wang et al. [2] introduced the robust formulation, which analyses the eroded, dilated, and intermediate realisations of the solution and imposes the length scale in the intermediate solution when all design configurations share a consistent topology. While this is one of the most effective approaches to addressing the issue, it has several drawbacks. For instance, it requires analysing each realisation of the solution, and when the physical realisations do not share a consistent topology, the formulation fails to impose the minimum length scale.

Another method for controlling the minimum length scale relies on the skeleton procedure, which identifies the medial zone of structural elements. In this vein, several studies have been developed, such as that by Zhang et al. [3], whose main weakness lies in the fact that the structural skeleton does not influence the sensitivity analysis. Additionally, the minimum length scale constraint may be incompatible with the selection of other optimisation parameters, thereby necessitating the adoption of a suboptimal topology.

This study introduces a novel methodology to address the minimum length scale constraint in optimisation problems. The primary objective is to ensure that this constraint does not require an excessive number of filters or additional variables that deviate from the design variable. Instead, it should maintain an intuitive geometric interpretation and be formulated in a straightforward manner, avoiding unnecessary complexity in sensitivity analysis.

To achieve this, two methods are proposed for integrating the constraint into the optimisation algorithm, which may also be employed complementarily if needed. The first

WIT Transactions on The Built Environment, Vol 216, © 2025 WIT Press
www.witpress.com, ISSN 1743-3509 (on-line)
doi:10.2495/HPSM250131

method is based on perimeter control, while the second introduces an innovative approach to obtaining a differentiable representation of the structure's skeleton.

2 PROBLEM STATEMENT

The proposed processes both require the cascade of variables illustrated in Fig. 1. The operation (1)→(3) describes the general density based topology optimisation procedure. The operations (2)→(4) and (2)→(5) describe the perimeter yielding steps using the grey-scale values and the gradient, P_ρ and P_∇, respectively. The transformation (4)→(6) produces the smooth perimeter $\tilde{P}$. Finally, the combined operation (4), (5)→(7) result in the skeletisation of the structure, S.

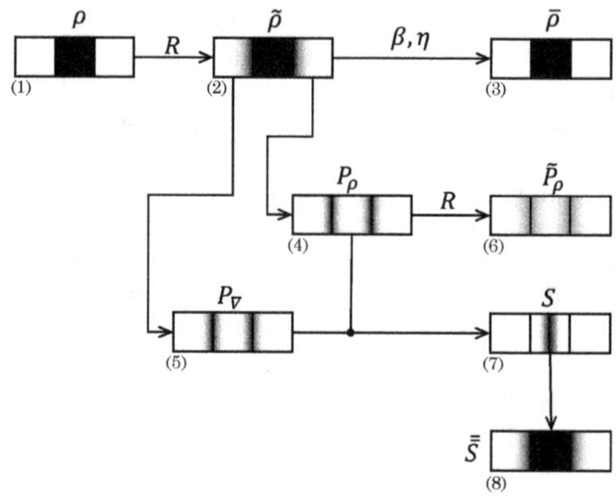

Figure 1: Cascade of variables.

2.1 Smoothing

Smoothing techniques are employed to prevent checkerboard patterns and control the minimum length-scale in density-based topology optimisation approaches. This work resorts to smoothing as well but extends its application to achieve perimeter control and enable local perimeter control. We applied a PDE filter based on a Helmholtz differential equation, as proposed by Lazarov and Sigmund [4], with consistent boundary conditions to avoid padding, following the approach of by Wallin et al. [5]. The governing equation incorporates l_r as the penalisation parameter for length-scale control, defined as $r = 2\sqrt{3}l_r$. Additionally, ϕ and $\tilde{\phi}$ as the initial and filtered variables, respectively, while l_s serves as a penalisation parameter to prevent material from sticking to the contour. Finally, $\boldsymbol{n}$ denotes the normal contour vector.

$$\begin{cases} -l_r^2\nabla^2\tilde{\boldsymbol{\phi}} + \tilde{\boldsymbol{\phi}} = \boldsymbol{\phi} \quad / \quad \boldsymbol{\phi} \in \{\boldsymbol{\rho}, \boldsymbol{P}_\rho\} \\ l_r^2\nabla\tilde{\boldsymbol{\phi}} \cdot \boldsymbol{n} = -l_s\tilde{\boldsymbol{\phi}} \quad in\ \partial\Omega \end{cases}. \tag{1}$$

2.2 Projection

After smoothing, we apply a projection filter to eliminate grey-scale transitions between solid and void regions. Further detail can be found in Sigmund [6], [7]. In this work, we use the projection function proposed in Wang et al. [2]:

$$\overline{\boldsymbol{\phi}} = \frac{\tanh(\beta\eta)+\tan\left[\beta(\widetilde{\boldsymbol{\phi}}-\eta)\right]}{\tanh(\beta\eta)+\tan\left[\beta(1-\eta)\right]}, \tag{2}$$

where β controls the sharpness of the transition, approximating a unit step function, and η defines the projection threshold.

2.3 Perimeter restriction

The following method evaluates the local perimeter across the computational domain. This point-wise calculation is performed over the smoothed field $\widetilde{\boldsymbol{\rho}}$ such that:

$$\boldsymbol{P}_{\rho} = 4\widetilde{\boldsymbol{\rho}}(1-\widetilde{\boldsymbol{\rho}}). \tag{3}$$

This equation is illustrated in Fig. 2.

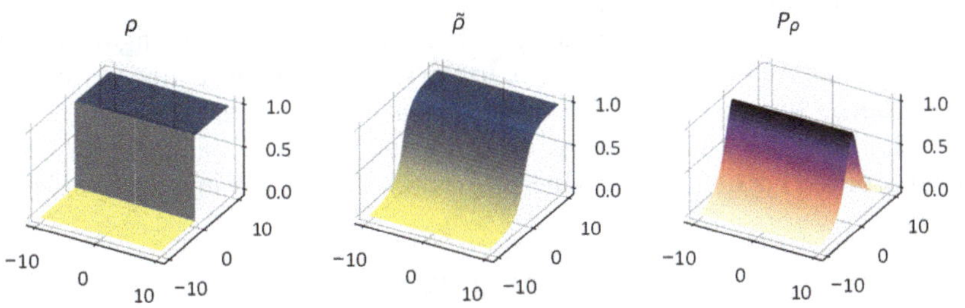

Figure 2: 1D example of the perimeter calculation.

The key idea behind this procedure resides in the fact that the filtered perimeter $\widetilde{\boldsymbol{P}}_{\rho}$ reaches a minimum value of $2/3$ at $x = 0$, except in cases where the structural element is thinner than the defined filter radius, in which case $\widetilde{\boldsymbol{P}}_{\rho} > 2/3$. Therefore, ensuring a filtered perimeter value of $2/3$ guarantees a minimum length scale of r.

Alternatively, this method is also effective for controlling the curvature of the contour at element junctions, as depicted in Fig. 3.

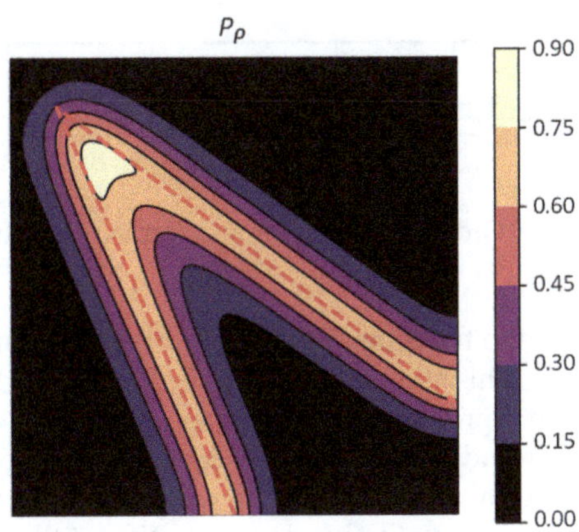

Figure 3: Perimeter at a junction.

2.4 The skeleton method

The skeleton can be defined as a continuous pixel-width representation of the structure's topology by reducing it to its central axis. Such graphical representation favours the evaluation of the design at the centre of its structural elements where we can easily identify problematic regions. One novelty of this work is the introduction of a derivable skeletisation process built upon simple filtering techniques.

The computation of the skeleton uses a function that becomes singular at the axis of the structural elements. That is possible due to the similarity between the perimeter function (3) and the gradient function (4), that differ only at the axis of the bars where the gradient is null.

The skeleton can be defined as a continuous, pixel-width representation of the structure's topology, reducing it to its central axis. This graphical representation facilitates the evaluation of the design at the centre of structural elements, allowing for the easy identification problematic regions. A key novelty of this work is the introduction of a differentiable skeletisation process built upon simple filtering techniques.

The skeleton is computed using a function that becomes singular along the central axis of the structural elements. This is possible due to the similarity between the perimeter function (3) and the gradient function (4), which differ only at the axis of the bars, where the gradient is null. By dividing these two functions, as expressed in eqn (5), a singularity is generated along the axis of the bars. After applying projection, this results in a precise description of the skeleton.

$$\boldsymbol{P}_{\nabla} = 2l_{\mathrm{r}}|\boldsymbol{\nabla}\tilde{\boldsymbol{\rho}}|. \tag{4}$$

$$s = \frac{\boldsymbol{P}_{\rho}}{2\boldsymbol{P}_{\nabla}} - 1. \tag{5}$$

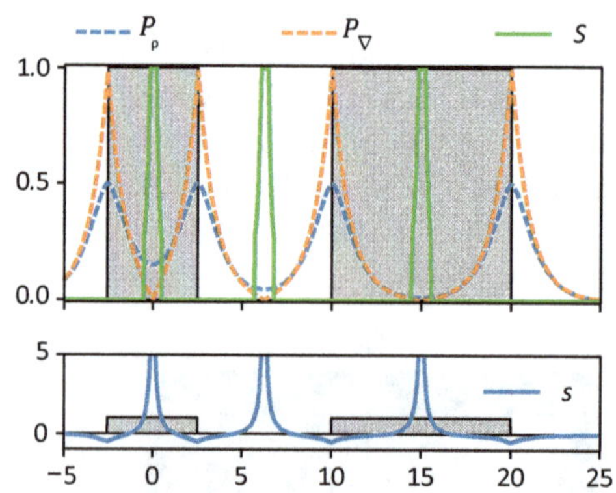

Figure 4: 1D description of the skeletisation process.

It is important to note that the function s in eqn (5) is independent of the width of solid or void elements near to the axis. This allows its influence to be isolated within a specific region, enabling an accurate skeleton representation of the topology.

A practical example is provided in Fig. 5, where the image on the left illustrates the classical MBB benchmark example, while the image on the right presents its skeletonised description. This result serves as a qualitative demonstration of the precision of the proposed method.

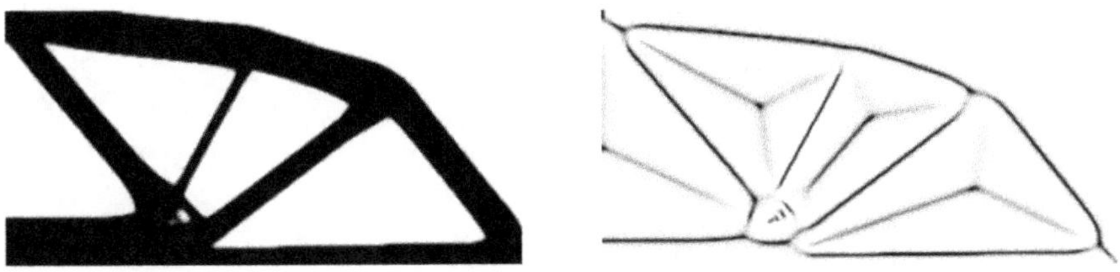

Figure 5: Skeletonised representation of a benchmark MBB structure.

3 OPTIMISATION PROBLEM

We consider the general topology optimisation problem described in Fig. 6. The objective is to determine the material distribution that results in the best response to the objective function F while satisfying the predefined design constraints:

$$\begin{array}{ll} \min(\boldsymbol{\rho}) & F(\boldsymbol{u}(\boldsymbol{\rho}), \boldsymbol{\rho}) \\ \text{st} & g(\boldsymbol{\rho}) \leq 0 \\ & l_p(\boldsymbol{\rho}) \leq 0 \\ & l_s(\boldsymbol{\rho}) \leq 0 \\ & l_v(\boldsymbol{\rho}) \leq 0 \\ & \rho_e \epsilon [0,1], \forall e \end{array} \tag{6}$$

In this formulation, $\boldsymbol{\rho}$ is the vector of independent design variables, and $\boldsymbol{u}$ denotes the nodal displacement vector. The functions $l_p(\boldsymbol{\rho})$, $l_s(\boldsymbol{\rho})$, and $l_v(\boldsymbol{\rho})$ impose restrictions on the local perimeter and the skeleton-based restrictions for the void and solid phases, respectively. The function $g(\boldsymbol{\rho})$ governs the total material volume and is defined as:

$$g(\boldsymbol{\rho}) = \frac{V(\boldsymbol{\rho})}{\alpha V_0} - 1 \leq 0, \tag{7}$$

where $V(\boldsymbol{\rho})$ is the current material volume, V_0 is the volume of the design domain, and α is the predefined maximum volume fraction.

Given that all constraints are differentiable with respect to the design variables $\boldsymbol{\rho}$, and assuming that the first-order derivative of the objective function $\boldsymbol{F}$ is available, this optimisation problem can be solved using the method of moving asymptotes (MMA) proposed by Svanberg [8].

3.1 Local perimeter constraint

The local perimeter constraint governs both the width of the structural members and the curvature at the junction regions. As introduced in Section 2.3, the local perimeter must satisfy the following condition:

$$\tilde{\boldsymbol{P}}_\rho > 2/3. \tag{8}$$

A valid constraint function is constructed using constraint aggregation with the *MellowMax* function, as proposed in Asadi and Littman [9]:

$$l_p(\boldsymbol{\rho}) = \frac{3}{2p_p} \log\left(\frac{1}{N_e} \sum_{e=1}^{N_e} \exp\left(p_p \cdot \tilde{P}_{\rho e}\right)\right) - 1 \leq 0. \tag{9}$$

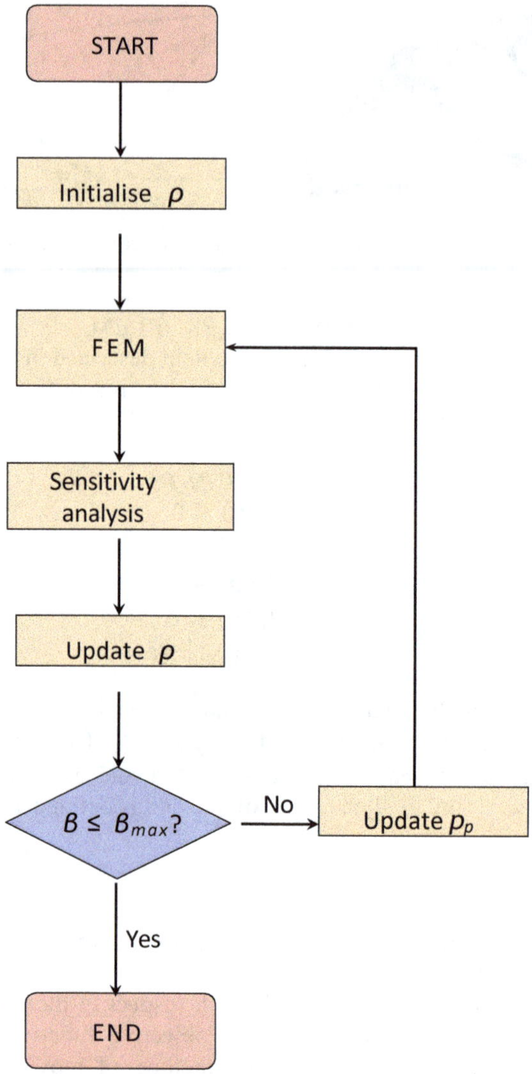

Figure 6: Flowchart of the topology optimization problem.

In eqn (9), $p_p \in (-\infty, \infty)$ and N_e is the number of independent design variables. Notably, as $p_p \to \infty$, eqn (9) converges to a *max* function. When $p_p \to 0$, it approximates the arithmetic mean, and as $p_p \to -\infty$ it behaves as a *min* function.

3.2 Skeleton-based constraint

The skeleton-based constraint is analogous to the local perimeter constraint, with the corresponding functions defined as:

$$l_s(\boldsymbol{\rho}) = \frac{1}{l_{lim}} \left[\frac{1}{N_e} \sum_{e=1}^{N_e} {l_{se}}^{p_s}\right]^{\frac{1}{p_s}} - 1 \le 0. \tag{10}$$

$$l_v(\boldsymbol{\rho}) = \frac{1}{l_{lim}}\left[\frac{1}{N_e}\sum_{e=1}^{N_e} l_{ve}{}^{p_v}\right]^{\frac{1}{p_v}} - 1 \leq 0. \tag{11}$$

In these equations, subscripts s and v refer to the skeleton constraint applied to the solid and void materials, respectively. The parameter l_{lim} represents their limit value, which is set to $l_{lim} = 0.9$.

4 NUMERICAL EXAMPLE

This example addresses the benchmark MBB optimisation problem, formulated as:

$$\begin{aligned} \min(\boldsymbol{\rho}) \quad & c = \boldsymbol{u}^T\boldsymbol{K}\boldsymbol{u} \\ \text{st} \quad & \boldsymbol{K}\boldsymbol{u} = \boldsymbol{f} \\ & g(\boldsymbol{\rho}) \leq 0 \\ & l_p(\boldsymbol{\rho}) \leq 0 \\ & l_s(\boldsymbol{\rho}) \leq 0 \\ & l_v(\boldsymbol{\rho}) \leq 0 \\ & \rho_e \epsilon [0,1], \forall e \end{aligned} \tag{12}$$

where $\boldsymbol{K}$ is the global stiffness matrix and $\boldsymbol{f}$ the nodal force vector.

The material properties are interpolated using the solid isotropic material with penalisation (SIMP) method:

$$\boldsymbol{K}(\boldsymbol{\rho}) = [E_{min} + (E_o + E_{min})\bar{\rho}_i{}^p]\boldsymbol{K_o}. \tag{13}$$

where p is the penalisation parameter, E_o is the Young's modulus of the solid material, E_{min} the Young's modulus of the void material, and $\boldsymbol{K_o}$ is the stiffness matrix of the isotropic material with unit Young's modulus.

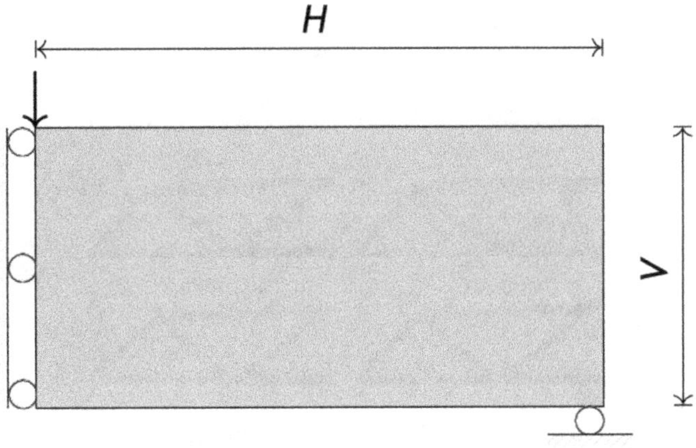

Figure 7: Design domain.

Fig. 8 presents an example of an MBB beam with a volume fraction of $v = 0.3$ and three different mesh densities. In all three cases, the design satisfies the minimum length scale requirements for both solids and voids regions, leading to highly consistent designs regardless of the mesh density.

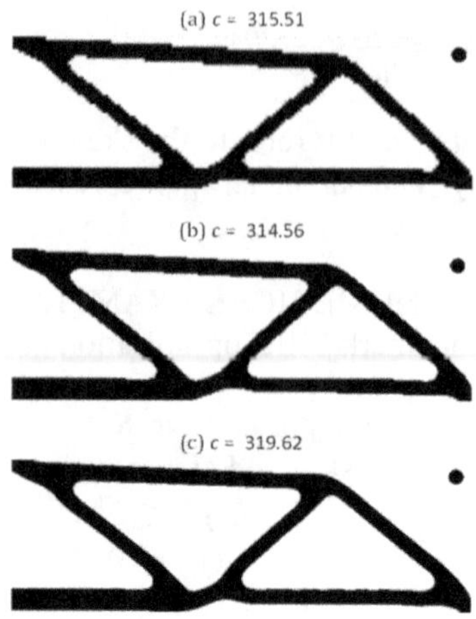

Figure 8: MBB results with different mesh densities: 150 × 50, 300 × 100 and 600 × 200, with filter radii being $r_{min} = 5, 10, 20$.

Fig. 9 illustrates the iteration process for the case depicted in Fig. 8(c). It is evident that the perimeter constraint not only enforces a minimum length scale by widening structural elements but also prioritises the most relevant bars. This approach removes material from non-essential components and, in some cases, eliminates them.

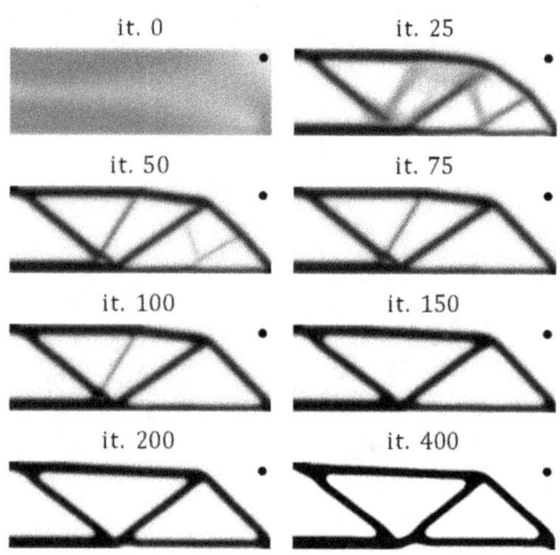

Figure 9: Design evolution.

This method proves particularly effective in scenarios with very stringent volume constraints. A constraint based solely on skeleton functions tends to increase the size of structural elements and may fail to produce a topology that aligns with the imposed volume limitation. This issue makes it challenging to achieve results similar to those shown in Fig. 8, where both the filter radius and the target volume are extremely small.

When the filter radius is relatively small compared to the design space, highly intricate topologies often emerge, characterised by numerous solid and void elements. However, the proposed method effectively simplifies the resulting topology while maintaining compatibility with the imposed volume constraint.

5 CONCLUSIONS

This work introduces a methodology for enforcing minimum size constraints in topology optimisation by leveraging local perimeter control and skeletal representation. These complementary approaches effectively maintain a prescribed length scale without requiring excessive filtering or additional finite element analyses. Perimeter control ensures that structural elements adhere to minimum size requirements, while the skeleton-based method identifies and resolves geometric inconsistencies, allowing the optimisation process to restart with an open design when necessary.

Additionally, perimeter control facilitates topology modifications by prioritising relevant solids and voids, effectively eliminating non-essential elements. Moreover, local perimeter control enables fine-tuned curvature adjustments, helping to smooth sharp edges. Numerical results in linear elasticity confirm the robustness and versatility of the proposed techniques. Furthermore, the method consistently achieves optimised designs, even under strict volume constraints, by implementing the necessary structural simplifications.

ACKNOWLEDGEMENTS

Grant I+D+i PID2019-109769RB-I00 funded by MCIN/AEI/10.13039/501100011033 and by 'ERDF A way of making Europe'. This project has received funding from the European Union's Horizon 2020 Joint Technology Initiative Shift2Rail under grant agreement No 101012456 (IN2TRACK3). The Basque Government also supported this work through IT-1764-22 – GIC21/191 and Elkartek22/69 projects.

REFERENCES

[1] Guest, J.K., Prévost, J.H. & Belytschko, T., Achieving minimum length scale in topology optimization using nodal design variables and projection functions. *Int. J. Numer. Methods Eng.*, **61**(2), pp. 238–254, 2004. https://doi.org/10.1002/nme.1064.

[2] Wang, F., Lazarov, B.S. & Sigmund, O., On projection methods, convergence and robust formulations in topology optimization. *Structural and Multidisciplinary Optimization*, **43**(6), pp. 767–784, 2011. https://doi.org/10.1007/s00158-010-0602-y.

[3] Zhang, W., Zhong, W. & Guo, X., An explicit length scale control approach in SIMP-based topology optimization. *Comput. Methods Appl. Mech. Eng.*, **282**, pp. 71–86, 2014. https://doi.org/10.1016/j.cma.2014.08.027.

[4] Lazarov, B.S. & Sigmund, O., Filters in topology optimization based on Helmholtz-type differential equations. *Int. J. Numer. Methods Eng.*, **86**(6), pp. 765–781, 2011. https://doi.org/10.1002/nme.3072.

[5] Wallin, M., Ivarsson, N., Amir, O. & Tortorelli, D., Consistent boundary conditions for PDE filter regularization in topology optimization. *Structural and Multidisciplinary Optimization*, **62**(3), pp. 1299–1311, 2020. https://doi.org/10.1007/s00158-020-02556-w.

[6] Sigmund, O., Manufacturing tolerant topology optimization. *Acta Mechanica Sinica*, **25**(2), pp. 227–239, 2009. https://doi.org/10.1007/s10409-009-0240-z.

[7] Sigmund, O., Morphology-based black and white filters for topology optimization. *Structural and Multidisciplinary Optimization*, 33(4–5), pp. 401–424, 2007. https://doi.org/10.1007/s00158-006-0087-x.

[8] Svanberg, K., The method of moving asymptotes: A new method for structural optimization. *Int. J. Numer. Methods Eng.*, **24**(2), pp. 359–373, 1987. https://doi.org/10.1002/nme.1620240207.
[9] Asadi, K. & Littman, M.L., An Alternative Softmax operator for reinforcement learning. *ICML'17: Proceedings of the 34th International Conference on Machine Learning*, pp. 243–252, 2017.

SECTION 8
MULTI-OBJECTIVE OPTIMISATION

MULTI-ATTRIBUTE DECISION-MAKING IN PRESTRESSED CONCRETE ROAD FLYOVER DESIGN

LORENA YEPES-BELLVER[1], JULIÁN ALCALÁ[2] & VÍCTOR YEPES[2]
[1]Mechanics of Continuous Media and Theory of Structures Department, Universitat Politècnica de València, Spain
[2]Institute of Concrete Science and Technology (ICITECH), Universitat Politècnica de València, Spain

ABSTRACT
Economic and environmental factors are essential considerations in bridge design. This study aims to evaluate the design criteria for a prestressed concrete road overpass, seeking an optimal balance among cost, carbon emissions and embedded energy. A Kriging surrogate model generates a response surface, extracting a Pareto front to identify optimal solutions for each objective function. Multi-criteria decision-making methods, including the weighting-sum and multi-attribute approaches, are applied across multiple scenarios to determine the most effective compromise solutions. Based on these results, preliminary design guidelines are proposed for this type of structure. The findings reveal that compromise decks exhibit a depth-to-main span ratio close to 1/30. Designs that balance cost, emissions, and energy consumption or prioritize environmental factors tend to produce slender decks. In contrast, cost-driven optimization results in greater deck depths. This research systematically integrates economic and environmental criteria in bridge design, offering practical insights for achieving sustainable and efficient structural solutions.
Keywords: multi-criteria decision-making, prestressed concrete, sustainability, Kriging model, slab bridges, optimization.

1 INTRODUCTION

The Sustainable Development Goals (SDGs) in Agenda 2030 call for maximizing benefits and minimizing negative economic, environmental, and societal impacts. SDG 9 (industry, innovation, infrastructure) highlights the need for sustainable infrastructure by 2030, given the construction sector's economic and environmental impact [1]–[3].

Optimizing structural design has advanced low-carbon cement research [4], [5]. Tiwari et al. [6] examined CO_2 reduction costs in construction and its impact on employment and materials. Itoh and Kitagawa [7] proposed a life-cycle assessment for CO_2 emissions in new bridges. Sharma et al. [8] found that building operations generate over 50% of total emissions.

Bridges are crucial for connectivity, making sustainability essential for safety and service. Incorporating design criteria is essential to reducing embedded CO_2 in reinforced concrete structures. Sustainability relies on economic, social, and environmental pillars, yet social sustainability remains less explored [9]. Achieving balance requires decision-making strategies for selecting bridge solutions based on lifecycle criteria [10]. Balali et al. [11] classified decision-support systems into three bridge lifecycle phases: planning and design, construction, and operation and maintenance, a classification supported by others [12], [13].

The research assesses bridge sustainability across construction and maintenance. Economic sustainability optimizes construction [14] and maintenance costs [15], [16]. Environmental assessments explore design geometry [17], [18] and maintenance strategies [19], [20]. Hammervold et al. [21] analyse the lifecycle of bridges, including routine maintenance. Zhang et al. [22] assess uncertainties in environmental impact evaluation.

Only recently has research addressed the social impacts of the bridge [23], [24] and road [25] construction and maintenance. Gervásio and Simões da Silva [26] assess social aspects by incorporating transport externalities. Sustainability evaluation requires assessing each

WIT Transactions on The Built Environment, Vol 216, © 2025 WIT Press
www.witpress.com, ISSN 1743-3509 (on-line)
doi:10.2495/HPSM250141

dimension individually and collectively. Since sustainability criteria often conflict [27], multi-criteria decision-making (MCDM) methods are essential tools [28]. MCDM techniques, widely used in engineering, integrate stakeholder preferences and manage uncertainties [29], [30]. Other reviews highlight their use in civil engineering, although applications for infrastructure sustainability are still emerging [31]–[33].

Reviews on MCDM applications in sustainable infrastructure have been noted, and the growth of research since 2015 is in line with the SDGs [34], [35]. However, the application of MCDM to bridge design remains limited [36]. This paper reviews MCDM use in sustainable bridge design, examining sustainability assessments and identifying knowledge gaps to achieve SDG 9.

Decision-making involves various methods and objectives. MCDM is classified into multiple attribute decision-making (MADM) and multiple objective decision-making (MODM) [37]. MADM, an a priori process, ranks discrete variables through expert weightings. MODM, an a posteriori process, generates a Pareto front of optimal solutions for expert selection.

Traditional methods often fail to address real-world complexities [38]. Sustainable bridge design requires handling vast information and diverse perspectives. Hybrid MCDM models offer a promising solution [39].

Unlike single-objective optimization [40], [41], multi-objective optimization identifies decision variables that satisfy constraints while optimizing multiple conflicting objectives [42]. The resulting models encapsulate performance criteria that often oppose each other, requiring decision-makers to navigate trade-offs where improvements in one objective typically lead to suboptimal outcomes in another. Thus, sustainable bridge design involves identifying trade-offs that provide balanced, acceptable values across all objectives [43].

This study aims to contribute to the current research on sustainable bridge design by assessing the dimensions of sustainability and employing MADM techniques.

2 MATERIALS AND METHODS

This section explores the multi-attribute (MA) approach derived from MCDM to create an indicator that integrates cost reduction, carbon emissions, and embedded energy criteria to determine a trade-off solution. The methodology identifies a trade-off between the objective functions of cost, CO_2 emissions, and embedded energy in bridge design. First, Latin hypercube sampling (LHS) generates 50 seed decks with uniform distribution across key variables, such as deck depth, base width, and concrete grade. Each design is analysed for reinforcement and associated objective function costs. Kriging is applied to generate response surfaces for each objective, providing accurate performance estimates. An additional 1,000 designs are created using LHS, with Kriging predicting their values. The Pareto frontier is identified by selecting non-dominated solutions from the complete set. An MCDM framework ranks the solutions, and the final optimal design is determined by integrating rankings from the different MCDM methods.

2.1 LHS

LHS ensures uniform sample distribution by stratifying each variable's range into equal intervals. A random value is chosen per interval, and variables are randomly paired to form input vectors. The sampling process determines the sample's number and location. Sample size correlates with the number of variables, with larger samples ensuring greater metamodel accuracy.

2.2 Kriging surrogate model

Metamodel use in structural optimization has risen, especially by Kriging [44]. Kriging predicts the value of a characteristic at a given point using a set of observed values. This approach enables output estimation without requiring extensive structural analysis. As a deterministic surrogate model, Kriging ensures consistent results for identical inputs, as it does not incorporate random error. Eqn (1) expresses the formulation of a Kriging in general:

$$y(x) = f(x) + Z(x), \tag{1}$$

where $Z(x)$ is a stochastic, zero-mean spatially correlated residual modelled using a covariance function, and $f(x)$ is the approximation function similar to a regression model. It comprises the deterministic trend component, commonly represented by eqn (2):

$$f(x) = \sum_{i=1}^{n} \beta_i f_i(x). \tag{2}$$

Here, $f_i(x)$ represents the known basis functions, while β_i are unknown coefficients that define the trend basis or mean. This model is deterministic, implying that its response exhibits no random error. Consequently, the output remains consistent when the same input parameters are applied repeatedly.

2.3 The multi-attribute approach

The multi-attribute (MA) approach is a mathematical framework with axioms designed to analyse decisions involving multiple competing outcomes. This study applies the procedure proposed by Giustozzi et al. [45]. The approach represents the rescaled values of sustainability criteria (cost, CO_2 emissions, and embodied energy) in a three-dimensional model. The suitability of each alternative is assessed based on the volume within this representation. This study suggests using eqn (3) to calculate this volume, incorporating weights for each criterion. The point (1, 1, 1) represents the best solution across all three criteria, corresponding to a volume of 1.

$$MA_i = \prod_{j=1}^{n} \left(r_{ij}\right)^{w_j}, \qquad \prod_{j=1}^{n} w_j = 1 \ \text{ and } \ w_j > 0 \ \ \forall j, \qquad i = 1, \ldots, m. \tag{3}$$

3 CASE STUDY

3.1 Post-tensioned road flyover description

This study focuses on optimizing voided slab decks for in situ post-tensioned road overpasses with spans of 24–34–28 m, totalling 86 m long. The design employs a slab with a constant depth and a straight profile, featuring a deck width of 8.30 m that accommodates two 3.50 m lanes, 0.65 m barriers on both sides and an integrated pedestal. This configuration is commonly implemented over two-lane dual carriageways and enhances structural efficiency (see Fig. 1).

The bridge is at kilometre 441 of the A-7 motorway in Cocentaina, Alicante. The structure features a base width of 4.00 m, a depth of 1.35 m, and a lateral cantilever of 1.75 m. Key dimensions include a = 0.20 m, b = 0.10 m and d = 0.40 m. The internal voids consist of four circular cylinders, each with a diameter of 0.60 m, yielding a void volume of 0.14 m^3/m^2. Fig. 2 shows the overpass.

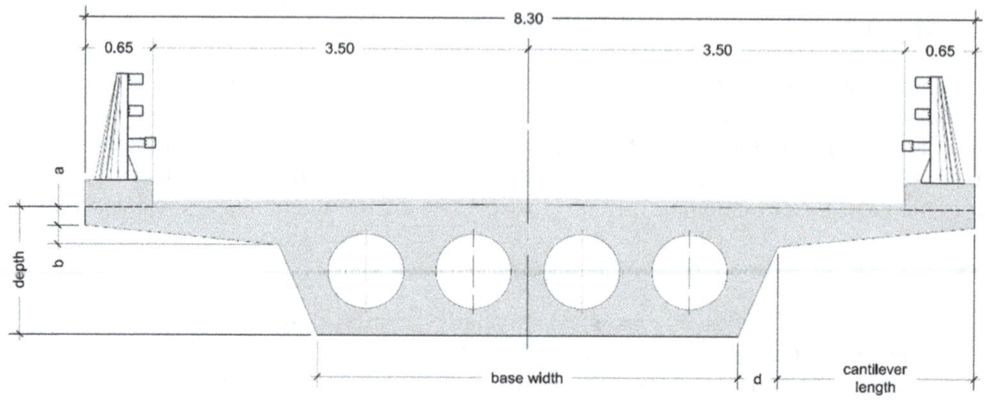

Figure 1: Cross-sectional view of the light-weighted gull wing PC slab bridge deck.

Figure 2: Overpass at kilometre 441 of the A-7 highway in Cocentaina. *(Source: L. Yepes-Bellver.)*

The passive steel B-500S includes longitudinal and transverse reinforcement. Longitudinal bars form the deck's base, enhancing bending and torsion resistance in key areas. Reinforcement covers 70% of the span and 15% above piers.

Structural constraints comply with Eurocode 2 serviceability and ultimate limit states, considering Eurocode 1 actions, including 44 kN/m dead loads and exposure class XC4. Checks cover bending, vertical and longitudinal shear, torsion, combined effects, compression, tension, cracking, vibration, constructability, and geometry.

This study used CSiBridge 21.0.0 [46] to develop a three-dimensional structural analysis and dimensioning model. Various design alternatives were assessed to evaluate sectional stresses obtained as sectional forces from the model under applied loads. The sectional design approach was utilized to identify the stress resultants resisted by each section. Acting and resisting stresses were analysed individually, aligning with force calculations in [47]–[49].

3.2 Objective functions

Each slab deck construction entails economic and sustainability-related costs, including CO_2 emissions and embedded energy associated with concrete grade, steel quantity, formwork, and lighting. These factors were analysed using BEDEC unit costs. This widely recognized database, extensively used by national contractors, provides a representative and realistic cost analysis. These variations do not affect the methodology's applicability. Unit costs are weights in the objective function, multiplying each element's measurements. Table 1 presents costs for cost, emissions, and energy consumption [47]–[49].

Table 1: Economic and sustainable unit costs of the deck [47]–[49].

Deck unit	Unit	Cost (EUR)	CO_2 (kg)	Energy cost (MW·h)
C-30 concrete	m^3	99.81	227.01	596.91
C-35 concrete	m^3	104.57	263.96	612.22
C-40 concrete	m^3	109.33	298.57	646.61
C-45 concrete	m^3	114.10	330.25	681.00
C-50 concrete	m^3	118.87	358.97	715.39
Steel reinforcement	kg	1.16	3.03	10.44
Steel prestressed	kg	3.40	5.64	12.99
Voids	m^3	99.81	604.42	1137.50
Formwork	m^2	33.81	2.24	8.70

4 RESULTS AND DISCUSSION

4.1 Sampling and data collection

LHS enhances space exploration to identify locally optimal solutions. The analysed variables include concrete compressive strength (30–50 MPa), slab depth (1.15–1.70 m), and base dimensions (3.00–5.00 m in 5 cm increments). A sample size of 30 individuals has yielded favourable results in concrete structures [47], [48]. This study used 50 decks, called 'seed' solutions, to generate the response surface through Kriging. According to previous work, this sample includes the optimal solutions for each objective function – cost, carbon emissions, and embedded energy [47]–[49]. Table 2 shows the values of the sample of 50 decks and the assessment of their objective functions.

These seed solutions formed the basis for a response surface using a Kriging surrogate model, which generated 1,000 additional solutions evaluated for three objective functions. The 50 seed decks are actual bridges with accurate assessments, while the 1,000 generated solutions are Kriging model predictions. The analysis identified the Pareto front from all 1,050 solutions. Table 3 lists the Pareto front points, where the first four decks (#2, #6, #10, and #13) are seed solutions, and the remaining eighteen are predicted.

The lowest cost solution is #185, the lowest emissions is #475, and #13 is the lowest embodied energy. However, it is necessary to determine if a trade-off solution offers improvements across all three criteria.

Next, MCDM is applied to identify trade-off solutions among the non-dominated alternatives. Furthermore, various weighting scenarios are defined by applying MA. These frameworks enable a comprehensive evaluation of the criteria weights, ensuring robustness in the decision process.

Table 2: Sampling of 50 decks and assessment of their objective functions.

Deck	Deck depth (m)	Base width (m)	Concrete grade (MPa)	Cost (EUR × 10^3)	CO_2 (kg × 10^3)	Energy (MW·h)
1	1.05	3.50	35	199.6	385.9	1237.9
2	1.10	3.40	35	181.5	386.5	1151.0
3	1.10	3.45	40	188.5	392.4	1124.2
4	1.10	3.80	45	194.4	417.3	1178.7
5	1.15	3.25	40	180.9	397.2	1054.7
6	1.15	3.35	40	180.2	395.5	1051.0
7	1.15	3.35	45	184.0	418.9	1074.8
8	1.15	3.40	35	189.7	411.1	1098.5
9	1.15	3.50	50	185.1	432.8	1116.5
10	1.15	3.55	40	190.1	391.4	1058.4
11	1.15	3.60	45	187.3	421.4	1104.8
12	1.15	3.65	45	187.5	422.9	1096.3
13	1.15	3.70	40	182.7	394.6	1038.3
14	1.15	3.90	35	188.3	416.8	1129.2
15	1.15	4.50	50	195.2	471.4	1199.6
16	1.20	3.35	45	182.2	418.5	1065.4
17	1.20	3.40	40	181.7	399.7	1059.9
18	1.20	3.60	40	184.6	400.9	1083.4
19	1.20	3.85	40	184.1	408.0	1065.9
20	1.20	4.85	50	196.8	472.4	1170.7
21	1.25	3.00	50	181.0	423.1	1101.0
22	1.25	3.35	35	181.8	398.6	1096.5
23	1.25	3.40	45	184.5	419.9	1084.9
24	1.25	3.50	45	181.3	420.5	1073.8
25	1.25	3.60	45	185.0	423.0	1078.7
26	1.25	4.70	40	194.7	462.9	1191.7
27	1.30	3.15	35	180.1	407.2	1106.9
28	1.30	4.45	30	195.2	436.7	1180.3
29	1.30	4.90	40	198.0	477.5	1215.8
30	1.35	3.60	35	186.8	416.6	1094.9
31	1.35	3.95	30	187.2	406.7	1103.2
32	1.35	4.25	45	194.4	455.4	1132.7
33	1.40	3.30	40	181.3	443.8	1152.3
34	1.40	3.45	45	186.1	470.0	1201.7
35	1.45	3.90	45	194.7	464.5	1145.2
36	1.45	4.35	35	192.1	439.4	1119.2
37	1.45	4.75	35	197.5	455.7	1163.6
38	1.50	3.35	45	190.5	455.4	1134.9
39	1.50	3.55	35	185.8	418.8	1105.4
40	1.50	4.05	45	193.3	482.7	1183.2
41	1.50	4.50	45	202.7	490.7	1189.5

Table 2: Continued.

Deck	Deck depth (m)	Base width (m)	Concrete grade (MPa)	Cost (EUR × 10^3)	CO_2 (kg × 10^3)	Energy (MW·h)
42	1.50	4.55	30	192.5	434.7	1138.0
43	1.55	3.20	30	182.4	404.0	1103.4
44	1.55	3.60	45	189.6	456.7	1140.8
45	1.55	4.10	35	205.8	460.4	1162.9
46	1.60	4.20	40	199.9	503.8	1267.9
47	1.65	3.45	45	191.8	464.1	1145.1
48	1.65	3.65	35	191.0	444.7	1149.9
49	1.70	3.80	45	208.5	484.9	1182.9
50	1.70	3.85	40	198.6	468.9	1165.5

Table 3: Pareto front decks and assessment of their objective functions.

Deck	Deck depth (m)	Base width (m)	Concrete grade (MPa)	Cost (EUR × 10^3)	CO_2 (kg × 10^3)	Energy (MW·h)
2	1.10	3.40	35	181.5	386.5	1151.0
6	1.15	3.35	40	180.2	395.5	1051.0
10	1.15	3.55	40	190.1	391.4	1058.4
13	1.15	3.70	40	182.7	394.6	1038.3
74	1.25	3.10	30	182.0	386.6	1113.5
108	1.43	3.08	35	177.9	413.8	1093.7
185	1.42	3.01	35	176.9	413.6	1099.3
199	1.37	3.03	35	177.2	410.4	1094.5
281	1.24	3.04	45	178.6	417.0	1083.0
296	1.22	3.05	45	179.1	417.1	1079.5
475	1.11	3.02	30	187.6	376.0	1167.3
514	1.34	3.23	30	181.3	394.3	1098.3
535	1.26	3.00	45	177.9	418.9	1086.7
570	1.19	3.08	35	181.5	395.1	1092.7
684	1.22	3.10	40	179.6	407.1	1077.6
691	1.35	3.09	40	178.0	420.8	1084.2
711	1.33	3.17	35	179.4	408.6	1093.0
730	1.27	3.02	40	177.7	414.6	1089.7
829	1.48	3.06	35	178.1	415.2	1083.8
903	1.25	3.06	35	179.5	395.6	1094.6
989	1.24	3.08	30	182.1	386.2	1114.7
1016	1.20	3.08	30	183.6	383.1	1110.8

The MA-0 scenario assigns equal weights to all criteria. In contrast, scenarios MA-1 to MA-6 follow a hierarchical weighting scheme, where the most important criterion is assigned twice the weight of the second, which in turn is allocated twice the weight of the third. These scenarios ensure that the product of the weights equals one, as defined by eqn (3). Table 4 presents the weights given to different scenarios for each criterion and the applied MA approach.

Table 4: Weights for different scenarios for each criterion and trade-off solution from the Pareto front applying MA.

MA	Cost	CO_2	Energy	Trade-off solution
MA-0	1.00	1.00	1.00	13
MA-1	2.00	1.00	0.50	6
MA-2	2.00	0.50	1.00	6
MA-3	1.00	2.00	0.50	1016
MA-4	1.00	0.50	2.00	13
MA-5	0.50	2.00	1.00	13
MA-6	0.50	1.00	2.00	13

Fig. 3 shows the MA scenarios. The surface is determined by the condition that the product of the weights for each criterion equals one. Solution #13 is preferred when the energy criterion weight is at least one-third, while solution #6 is preferred when the cost criterion takes precedence. Both have a concrete grade of 40 MPa and a deck depth of 1.15 m, differing only in base width: 3.35 m for deck #6 and 3.70 m for deck #13. The lowest-cost solution (#185) has a 23.5% greater deck depth, while the lowest-emission solution (#475) has a 3.5% smaller depth. Trade-off solutions (#6 and #13) have a higher concrete grade (40 MPa) than the lowest-cost (35 MPa) and lowest-emission (30 MPa) options. Additionally, base widths are smaller in the lowest-cost and lowest-emission solutions. The trade-off solutions feature a smaller deck depth and a slightly higher concrete grade than the lowest-cost and lowest-emission alternatives.

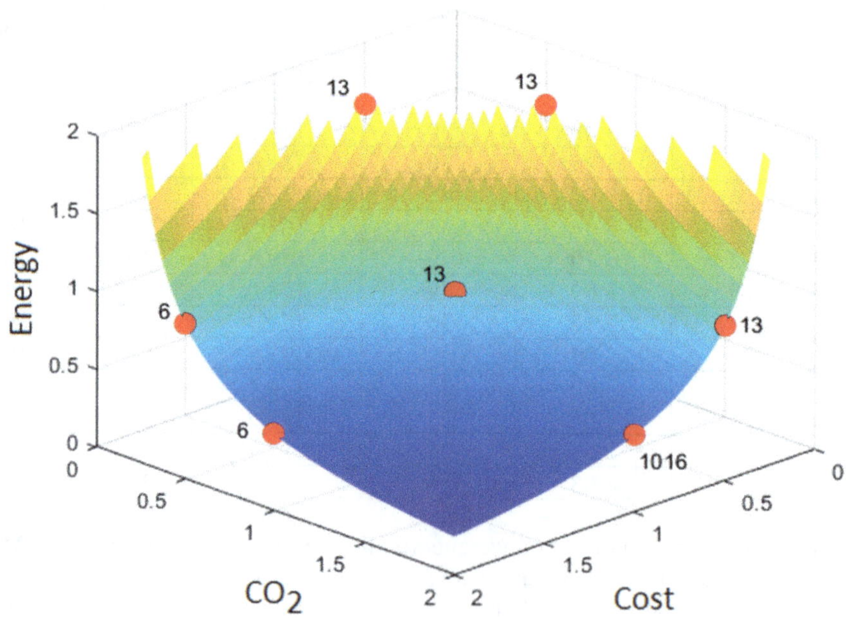

Figure 3: Preferred solution points for MA scenario.

The General Directorate of Public Roads of Spain (DGC) [50] suggests that the optimal ratio of deck depth to main span should fall between 1/22 and 1/30. At the same time, SETRA [51] recommends a ratio of 1/28 for three-span concrete decks with wide lateral cantilevers. In this study, decks #6 and #13 exhibit a deck depth-to-main span ratio close to 1/30. In contrast, deck #1016 is near 1/28 (Table 5).

Table 5: Measurements of the MA trade-off solutions.

Deck	Concrete grade (MPa)	Deck depth (m)	Span/deck depth	Concrete (m^3/m^2)	Passive steel (kg/m^2)	Active steel (kg/m^2)
6	40	1.15	29.57	0.56	71.25	16.48
13	40	1.15	29.57	0.61	69.41	16.65
1016	30	1.20	28.33	0.56	78.15	16.43

On the other hand, decks #6 and #13 employ a concrete grade of 40 MPa, higher than the 30 MPa used in deck #1016. However, when cost minimization is the primary objective, the lowest available concrete grade, 30 MPa, is the most economical option.

The DGC [49] recommends that for this type of bridge, the typical concrete volume ranges from 0.55 to 0.70 m^3 per m^2 of deck. Passive reinforcement generally falls between 70 and 100 kg per m^3 of concrete, while prestressed reinforcement ranges from 10 to 25 kg per m^3 of concrete, according to the trade-off solutions (Table 5).

5 CONCLUSIONS

The study employs a multi-attribute decision-making technique to facilitate the integration of multiple criteria into a cohesive decision framework, allowing for a comprehensive evaluation of alternatives based on their performance across economic and environmental dimensions. Using this methodology enhances the robustness of the decision-making process, ensuring that the selected solutions are well-aligned with sustainability goals.

The analysis of trade-off solutions has led to identifying key design criteria, which include maintaining a slab depth-to-span ratio close to 1/30, utilizing a concrete grade of 40 MPa, and optimizing the amount of passive and active reinforcement. These criteria contribute to structural efficiency and support the overall sustainability of the bridge design.

While the study identifies promising solutions, it acknowledges the need to validate the models under real-world conditions. Future research could expand the range of variables considered in the analysis to refine the optimization process further.

Practical recommendations include the importance of stakeholder engagement in the design process, ensuring that diverse perspectives are incorporated into decision-making. Additionally, the findings suggest that adopting a balanced approach to design, which prioritizes sustainability alongside cost considerations, can lead to more effective infrastructure solutions. This research supports the current discourse on sustainable bridge design and offers a framework for future studies to optimize infrastructure projects aligned with global sustainability goals, including social sustainability.

ACKNOWLEDGEMENTS

This work was supported by the Ministerio de Ciencia, Innovación y Universidades (Spain). Grant PID2023-150003OB-I00 funded by MCIN/AEI/10.13039/501100011033 and by 'ERDF A way of making Europe'.

REFERENCES

[1] United Nations, *Sustainable Development Goals*, United Nations. https://sdgs.un.org/goals.

[2] Schmidt, J.-S. & Osebold, R., Environmental management systems as a driver for sustainability: State of implementation, benefits and barriers in German construction companies. *Journal of Civil Engineering and Management*, **23**(1), pp. 150–162, 2017.

[3] Huang, G., Heokstra, A.Y., Krol, M.S., Galindo, A., Jägermeyr, J., Yu, C. & Wang, R., Water-saving agriculture can deliver deep water cuts for China. *Resources, Conservation and Recycling*, **154**, 104578, 2020.

[4] Gartner, E., Industrially interesting approaches to 'low-CO_2' cements. *Cement and Concrete Research*, **34**(9), pp. 1489–1498, 2004.

[5] Yang, L., Zmeureanu, R. & Rivard, H., Comparison of environmental impacts of two residential heating systems. *Building and Environment*, **43**(6), pp. 1072–1081, 2008.

[6] Tiwari, P., Parikh, J. & Sharma, V., Performance evaluation of cost-effective buildings: A cost, emissions and employment point of view. *Building and Environment*, **31**(1), pp. 75–90, 1996.

[7] Itoh, Y. & Kitagawa, T., Using CO_2 emission quantities in bridge life cycle analysis. *Engineering Structures*, **25**(5), pp. 565–577, 2003.

[8] Sharma, A., Saxena, A., Sethi, M., Shree, V. & Varun, M., Life cycle assessment of buildings: A review. *Renewable and Sustainable Energy Reviews*, **15**(1), pp. 871–875, 2011.

[9] Ostrom, E., A general framework for analyzing sustainability of social-ecological systems. *Science*, **325**(5939), pp. 419–422, 2009.

[10] Ozcan-Deniz, G. & Zhu, Y., A multi-objective decision-support model for selecting environmentally conscious highway construction methods. *Journal of Civil Engineering and Management*, **21**(6), pp. 733–747, 2015.

[11] Balali, V., Mottaghi, A., Shoghli, O. & Golabchi, M., Selection of appropriate material, construction technique, and structural system of bridges by use of multicriteria decision-making method. *Transp. Res. Rec. J. Transp. Res. Board*, **2431**, pp. 79–87, 2014.

[12] Malekly, H., Meysam Mousavi, S. & Hashemi, H., A fuzzy integrated methodology for evaluating conceptual bridge design. *Expert Syst. Appl.*, **37**, pp. 4910–4920, 2010.

[13] Dutil, Y., Rousse, D. & Quesada, G., Sustainable buildings: An ever-evolving target. *Sustainability*, **3**, pp. 443–464, 2011.

[14] Penadés-Plà, V., Martínez-Muñoz, D., García-Segura, T., Navarro, I.J. & Yepes, V., Environmental and social impact assessment of optimized post-tensioned concrete road bridges. *Sustainability*, **12**(10), p. 4265, 2020.

[15] Kabir, G., Sadiq, R. & Tesfamariam, S., A review of multi-criteria decision-making methods for infrastructure management. *Structure and Infrastructure Engineering*, **10**(9), pp. 1176–1210, 2014.

[16] Sabatino, S., Frangopol, D.M. & Dong, Y., Life cycle utility-informed maintenance planning based on lifetime functions: Optimum balancing of cost, failure consequences and performance benefit. *Structure and Infrastructure Engineering*, **12**(7), pp. 830–847, 2015.

[17] García-Segura, T., Yepes, V. & Frangopol, D.M., Multi-objective design of post-tensioned concrete road bridges using artificial neural networks. *Structural and Multidisciplinary Optimization*, **56**(1), pp. 139–150, 2017.

[18] Yepes, V., Martí, J.V. & García-Segura, T., Cost and CO_2 emission optimization of precast-prestressed concrete U-beam road bridges by a hybrid glowworm swarm algorithm. *Automation in Construction*, **49**, pp. 123–134, 2015.
[19] Navarro, I.J., Yepes, V. & Martí, J.V., Life cycle cost assessment of preventive strategies applied to prestressed concrete bridges exposed to chlorides. *Sustainability*, **10**(3), p. 845, 2018.
[20] Navarro, I.J., Martí, J.V. & Yepes, V., Reliability-based maintenance optimization of corrosion preventive designs under a life cycle perspective. *Environmental Impact Assessment Review*, **74**, pp. 23–34, 2019.
[21] Hammervold, J., Reenaas, M. & Brattebø, H., Environmental life cycle assessment of bridges. *Journal of Bridge Engineering*, **18**(2), pp. 153–161, 2013.
[22] Zhang, J., Huang, X., Fang, Y., Zhou, J., Zhang, H. & Li, J., Optimal inspection-based preventive maintenance policy for three-state mechanical components under competing failure modes. *Reliability Engineering and System Safety*, **152**, pp. 95–103, 2016.
[23] Bocchini, P., Frangopol, D.M., Ummenhofer, T. & Zinke, T., Resilience and sustainability of civil infrastructure: Toward a unified approach. *Journal of Infrastructure Systems*, **20**(2), 04014004, 2014.
[24] Navarro, I.J., Penadés-Plà, V., Martínez-Muñoz, D., Rempling, R. & Yepes, V., Life cycle sustainability assessment for multi-criteria decision making in bridge design: A review. *Journal of Civil Engineering and Management*, **26**(7), pp. 690–704, 2020.
[25] Sierra, L.A., Pellicer, E. & Yepes, V., Method for estimating the social sustainability of infrastructure projects. *Environmental Impact Assessment Review*, **65**, pp. 41–53, 2017.
[26] Gervásio, H. & Simões da Silva, L., A probabilistic decision-making approach for the sustainable assessment of infrastructures. *Expert Systems with Applications*, **39**(8), pp. 7121–7131, 2012.
[27] Kabir, G., Sadiq, R. & Tesfamariam, S., A review of multi-criteria decision-making methods for infrastructure management. *Structure and Infrastructure Engineering*, **10**(9), pp. 1176–1210, 2014.
[28] Cinelli, M., Coles, S.R. & Kirwan, K., Analysis of the potentials of multi criteria decision analysis methods to conduct sustainability assessment. *Ecological Indicators*, **46**, pp. 138–148, 2014.
[29] Zavadskas, E.K., Govindan, K., Antucheviciene, J. & Turskis, Z., Hybrid multiple criteria decision-making methods: A review of applications for sustainability issues. *Economic Research-Ekonomska Istraživanja*, **29**(1), pp. 857–887, 2016.
[30] Zavadskas, E.K., Antucheviciene, J., Turskis, Z. & Adeli, H., Hybrid multiple-criteria decision-making methods: A review of applications in engineering. *Scientia Iranica A*, **23**(1), pp. 1–20, 2016.
[31] Zavadskas, E.K., Antucheviciene, J. & Kaplinski, O., Multi-criteria decision making in civil engineering: Part I – A state-of-the-art survey. *Engineering Structures and Technologies*, 7(3), pp. 103–113, 2016.
[32] Zavadskas, E.K., Antucheviciene, J. & Kaplinski, O., Multi-criteria decision making in civil engineering: Part II – Applications. *Engineering Structures and Technologies*, 7(4), pp. 151–167, 2016.
[33] Jato-Espino, D., Castillo-Lopez, E., Rodriguez-Hernandez, J. & Canteras-Jordana, J.C., A review of application of multi-criteria decision making methods in construction. *Automation in Construction*, **45**, pp. 151–162, 2014.

[34] Navarro, I.J., Yepes, V. & Martí, J.V., A review of multicriteria assessment techniques applied to sustainable infrastructure design. *Advances in Civil Engineering*, **2019**, 6134803, 2019.
[35] Zavadskas, E.K., Antucheviciene, J., Vilutiene, T. & Adeli, H., Sustainable decision-making in civil engineering, construction and building technology. *Sustainability*, **10**(1), p. 14, 2018.
[36] Penadés-Plà, V., García-Segura, T., Martí, J.V. & Yepes, V., A review of multi-criteria decision-making methods applied to the sustainable bridge design. *Sustainability*, **8**(12), p. 1295, 2016.
[37] Hwang, C.L. & Yoon, K., *Multiple Attributes Decision Making: Methods and Applications*, Springer: Berlin, Germany, 1981.
[38] Liou, J.J.H. & Tzeng, G.-H., Comments on 'multiple criteria decision making (MCDM) methods in economics: An overview'. *Technol. Econ. Dev. Econ.*, **18**, pp. 672–695, 2012.
[39] Liou, J.J.H., New concepts and trends of MCDM for tomorrow: In honor of Professor Gwo-Hshiung Tzeng on the occasion of his 70th birthday. *Technol. Econ. Dev. Econ.*, **19**, pp. 367–375, 2013.
[40] Hernández, S., Fontan, A.N., Díaz, J. & Marcos, D., VTOP: An improved software for design optimization of prestressed concrete beams. *Advances in Engineering Software*, **41**(3), pp. 415–421, 2010.
[41] Carbonell, A., González-Vidosa, F. & Yepes, V., Design of reinforced concrete road vaults by heuristic optimization. *Advances in Engineering Software*, **42**(4), pp. 151–159, 2011.
[42] Pons, J.J., Villalba Sanchis, I., Insa Franco, R. & Yepes, V., Life cycle assessment of railway track substructures: Comparison of ballast and ballastless rail tracks. *Environmental Impact Assessment Review*, **85**, 106444, 2020.
[43] Osyczka, A., Multicriteria optimization for engineering design. *Des. Optim.*, (ed. J.S. Gero), Academic Press: New York, pp. 193–227, 1985.
[44] Cressie, N., The origins of Kriging. *Mathematical Geology*, **22**, pp. 239–252, 1990.
[45] Giustozzi, F., Crispino, M. & Flintsch, G., Multi-attribute life cycle assessment of preventive maintenance treatments on road pavements for achieving environmental sustainability. *Int. J. Life Cycle Assess.*, **17**(4), pp. 409–419, 2012.
[46] Computers and Structures, Inc., CSiBridge v21.0.0, Berkeley, CA, USA, 2018.
[47] Yepes-Bellver, L., Brun-Izquierdo, A., Alcalá, J. & Yepes, V., CO_2-optimization of post-tensioned concrete slab-bridge decks using surrogate modeling. *Materials*, **15**(14), p. 4776, 2022.
[48] Yepes-Bellver, L., Brun-Izquierdo, A., Alcalá, J. & Yepes, V., Embodied energy optimization of prestressed concrete road flyovers by a two-phase Kriging surrogate model. *Materials*, **16**(20), p. 6767, 2023.
[49] Yepes-Bellver, L., Brun-Izquierdo, A., Alcalá, J. & Yepes, V., Surrogate-assisted cost optimization for post-tensioned concrete slab bridges. *Infrastructures*, **10**(2), p. 43, 2025.
[50] Dirección General de Carreteras, Obras de Paso de Nueva Construcción: Conceptos Generales, Ministerio de Fomento, Centro de Publicaciones, Madrid, Spain, 2000.
[51] SETRA, Ponts-Dalles. Guide de Conception, Ministère de l'Equipement, du Logement des Transports et de la Mer, Bagneux, France, 1989.

SHAPE OF THE PARETO FRONT AND SENSITIVITY ANALYSIS AT THE PARETO POINTS IN MULTI-OBJECTIVE OPTIMIZATION

SANTIAGO HERNÁNDEZ & FÉLIX NIETO
School of Civil Engineering, University of Coruña, Campus de Elviña, Spain

ABSTRACT
Multi-objective optimization is a formulation of engineering design that considers several properties of the prototype to be optimized where there is conflict amongst them. Namely, increasing the quality of one produces worsening in others. This is why a surface coined Pareto front is created, composed of Pareto points that are trade-off values of the objective functions. Multi-objective optimization algorithms usually transform the initial formulation in a single objective optimization with an objective function that is a combination of the complete set. Each process produces a Pareto point, thus reducing the number of single optimizations and increasing the information on the Pareto points are very important metrics for an algorithm. Two techniques with that aim are presented. One associated with the description of the unconstrained and constrained Pareto points and how they shape the Pareto front. The other describes a new type of sensitivity analysis at the Pareto points, namely the sensitivity between objective functions, the sensitivity analysis of the design variables in relation to the objective functions. Two application examples are presented, namely portal frames with two objective functions (volume of material and flexibility) subject to stress constraints. One has the Pareto front composed of unconstrained Pareto points while the other is almost entirely composed of constrained Pareto points.
Keywords: multi-objective optimization, Pareto front, sensitivity analysis, structural design, constrained and unconstrained Pareto points.

1 INTRODUCTION

Multi-objective optimization is a quite common formulation in structural and multidisciplinary optimization. Many designers want to improve more than one property of the design. Usual situations are the amount of material and stiffness of a structure, lift versus drag forces in aircraft design or available volume versus amount of material in multi-lobe fluid tanks. These situations lead to conflicting objective functions and this is why the concept of Pareto point appears, that is, a design with an appropriate trade-off between the objective functions. The complete set of Pareto points creates the Pareto front that visualizes the geometry of the Pareto points and is a useful tool for deciding the final design. Multi-objective optimization has been studied since many decades ago and several books [1], [2] describe this approach of numerical optimization; techniques as the weighting method [3], the goal and compromise programming [4], the min–max approach [5] of the physical programming [6] have been proposed to obtain Pareto points. In industry, multi-objective optimization has been applied to automotive [7] and aircraft engineering [8]–[10] or tall buildings design [11].

The procedure to obtain a Pareto point requires to solve a single objective optimization problem, thus it is very important to create methods that reduce the number of points necessary to generate the Pareto front and also techniques that provide as much information as possible at each Pareto point, otherwise the size of the problems considered may lead to high computation times making difficult to tackle multi-objective optimization.

The research shown in this paper is devoted to the mentioned topic. First, a discussion about the nature of Pareto points and how they create the shape of the Pareto front is presented, then a formulation of a first order sensitivity analysis at the Pareto points is

WIT Transactions on The Built Environment, Vol 216, © 2025 WIT Press
www.witpress.com, ISSN 1743-3509 (on-line)
doi:10.2495/HPSM250151

described indicating the classes of information that can be obtained and how the designer can take profit of them.

Two application examples of multi-objective optimization are used to demonstrate the capabilities and the advantages of the formulation presented.

2 CLASSES OF PARETO POINTS IN MULTI-OBJECTIVE OPTIMIZATION PROBLEMS

A multi-objective optimization problem is defined by a vector of design variables $\mathbf{X} = |x_1,\ldots,x_n|$, a set of constraints $g_j(\mathbf{X}) \leq 0 \quad (j = 1,\ldots,m)$ and some objective functions $F = |f_1,\ldots,f_i,\ldots,f_K|$.

The aim of the problem is to obtain the best values of each f_k (k = 1, …, K) but usually the objective functions conflict. Therefore, no single point $\mathbf{X}^*$ exists that provides the best values of all objective function, in other words, there is not a point $\mathbf{X}^*$ that gives $f_k(\mathbf{X}^*) = \text{best } f_k(\mathbf{X}) \quad (k = 1,\ldots K)$.

Because of that, a class of points called Pareto points is defined that accomplishes:

a) $f_k(\mathbf{X}) \leq f_k(\mathbf{X}^P) \qquad (k = 1,\ldots K)$

b) $f_k(\mathbf{X}) \leq f_k(\mathbf{X}^P)$,

for at least one objective function

Fig. 1 presents an example of multi-objective optimization problem with two objective functions and two design variables. Fig. 1(a) shows the feasible region in the design variables space and two realizations of each objective function. Fig. 1(b) shows the domain of solutions in the objective functions space and the line E–F of Pareto points that constitutes the Pareto front. The Pareto points can be constrained or unconstrained: constrained Pareto points correspond to values of the design variables located on the boundary of the feasible region and unconstrained Pareto points are those created by values of the design variables located inside the feasible region. Both classes of Pareto points appear in Fig. 1(c).

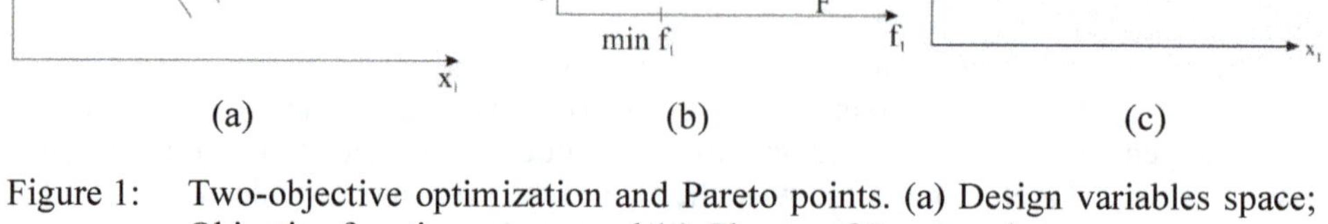

Figure 1: Two-objective optimization and Pareto points. (a) Design variables space; (b) Objective functions space; and (c) Classes of Pareto points.

Constrained Pareto points are defined as follows:

1. Values of the objective functions

$$f_k = f_k\left(\mathbf{X}^P\right) \qquad (k=1,...K) \tag{1a}$$

2. A set of active constraints

$$g_j\left(\mathbf{X}^P\right)=0 \qquad (j=1,...J) \tag{1b}$$

Unconstrained Pareto points are defined by:

1. Values of the objective functions

$$f_k = f_k\left(\mathbf{X}^P\right) \qquad (k=1,...K) \tag{2a}$$

2. The condition that the gradients of the objective functions have the same direction.

$$\nabla f_1 = -\mu \nabla f_2 \tag{2b}$$

Examples of constrained Pareto points are the broken lines of Fig. 2(a) while Fig. 2(b) shows the orientation of the gradient of the objective functions in an unconstrained Pareto point. Overall, a Pareto point can be defined as:

1. Values of the objective functions

$$f_k = f_k\left(\mathbf{X}^P\right) \qquad (k=1,...K) \tag{3a}$$

2. Set of conditions

$$\mathbf{C}_j\left(\mathbf{X}^P\right)=0 \qquad (j = 1, \ldots, n-1) \tag{3b}$$

where eqn (3b) corresponds to eqn (1b) in constrained Pareto points and to eqn (2b) in unconstrained Pareto points.

3 SENSITIVITY ANALYSIS IN MULTI-OBJECTIVE OPTIMIZATION

A usual approach to solve multi-objective optimization problems is to define a combined objective function using a vector $\boldsymbol{\lambda}$ of weights and creating the following problem:

$$\min F = \boldsymbol{\lambda}^T \mathbf{F} = \sum_{k=1}^{K} \lambda_k f_k \tag{4a}$$

subject to

$$g_j\left(\mathbf{X}\right) \le 0 \qquad (j=1,...m) \tag{4b}$$

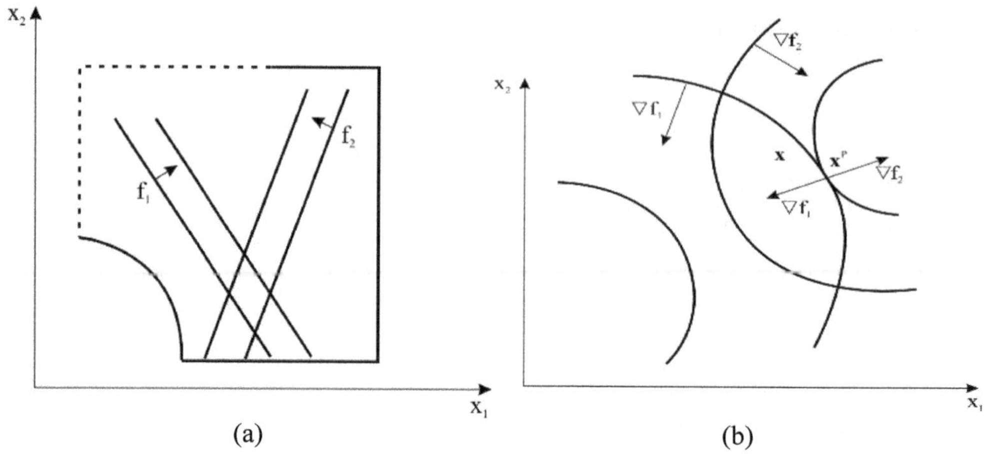

Figure 2: Classes of Pareto points. (a) Constrained Pareto points; and (b) Unconstrained Pareto points.

Usually the objective functions are normalized according to

$$\bar{f}_k = \frac{f_k - f_k^L}{f_k^U - f_k^L} \qquad (k = 1,...K) \tag{5}$$

to guarantee the same order of magnitude, f_k^L and f_k^U being the lower and upper values of each objective function f_k $(k = 1,...K)$. With each vector $\boldsymbol{\lambda}$, a Pareto point is obtained. Fig. 3 shows the flowchart of the process defined by expression (4), the Pareto points and the first derivative at a Pareto point.

Each time a Pareto point is obtained the designer evaluates how many points will be necessary to have an idea about the shape of the Pareto front and how much a design will differ from one Pareto points to another. Thus, two questions are relevant: How much f_2 changes on the Pareto front if f_1 changes and how much the vector of design variables $\mathbf{X}^p$ of a Pareto point varies if f_1 changes. The answers to these questions contribute to reducing the number of Pareto points necessary for taking decisions and to extract more information from each Pareto point. The way to do it is to calculate two classes of sensitivities at a Pareto point, specifically:

1. Sensitivities of an objective function with regards to the other ones, that inform of the rate of change of the objective functions at the Pareto points

$$df_K/df_k \quad (k = 1, \ldots, K-1) \tag{6}$$

2. Sensitivities of the design variables $\mathbf{X}$ with regard to the ob functions at a Pareto point $\mathbf{X}^p$

$$\frac{dx_i^P}{df_k} \quad (i = 1,...m) \quad (k = 1,...K) \tag{7}$$

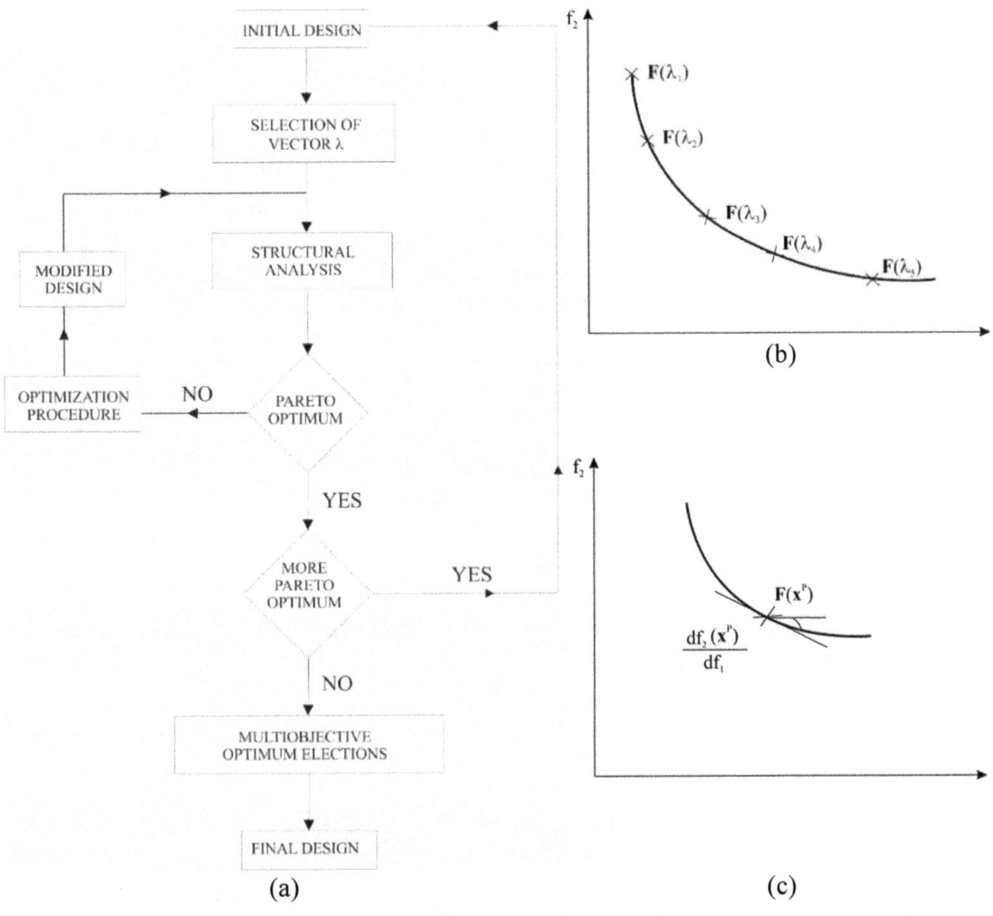

Figure 3: Multi-objective optimization and Pareto front. (a) Flowchart of multi-objective optimization; (b) Pareto points and Pareto front; and (c) First order derivative at a Pareto point.

The procedure to obtain the sensitivities of eqns (6) and (7) is as follows: Differentiating the objective function f_k at a Pareto point

$$df_k = \sum_{i=1}^{n} \frac{\partial f_k}{\partial x_i} dx_i \tag{8}$$

or

$$\sum_{i=1}^{n} \frac{\partial f_k}{\partial x_i} \frac{dx_i}{df_k} = 1 \tag{9}$$

Differentiating eqn (3b) at a Pareto point and dividing by df_k, provides

$$\sum_{i=1}^{n} \frac{\partial \mathbf{C}_j}{\partial x_i} \frac{dx_i}{df_k} = 0 \qquad (j = 1, \ldots n-1) \tag{10}$$

The combination of eqns (9) and (10) provides a system of linear equations in which the coefficients $\partial f_k/\partial x_i$ (i = 1, …, n) are known, the coefficients $\partial \mathbf{C}_j/\partial x_i$ (j = 1, …, n−1), (i = 1 …, n) are also known when gradient-based algorithms are used in the optimization, so the unknowns are dx_i/df_k (i = 1, …, n). The solution of this system of equations for each objective function f_k (k = 1, …, K−1) gives information to the designer about the variation of the design variables **X** in the neighbourhood of a Pareto point

$$x_i^{\Delta f_k} = x_i^P + \sum_{k=1}^{K-1} \frac{dx_i}{df_k} \Delta f_k \qquad (i = 1, \ldots, n) \tag{11}$$

Differentiating the objective function f_K and dividing by f_k

$$\frac{df_K}{df_k} = \sum_{i=1}^{n} \frac{\partial f_K}{\partial x_i} \frac{dx_i}{df_k} \qquad (k = 1, \ldots, K-1) \tag{12}$$

Therefore, the other class of sensitivity is easily obtained and informs on the variation of the objective function at a Pareto point

$$f_K^{\Delta f_k} = f_K^P + \sum_{k=1}^{K-1} \frac{df_K}{df_k} \Delta f_k \qquad (k = 1, \ldots, K-1) \tag{13}$$

4 APPLICATION EXAMPLES

Two examples of multi-objective optimization with two objective function and two design variables are presented to describe the performance of the methodology. It is important to consider that each example contains Pareto points of very different nature, being constrained in one case and mostly unconstrained in the other one.

4.1 Application example: Hinged portal frame

Fig. 4(a) shows a hinged frame subject to a horizontal force *F*. The design variables are the cross sections A_1, A_2 of the vertical and horizontal bars and the objective functions are the volume of material *V* and the lateral displacement *δ*. Stress constraints are the normal stress in the bars produced by the bending moment and the axial forces. The geometrical and mechanical data are [12]:

$$E = 2.07\cdot 10^5 \text{ MPa} \quad F = 2.07 \text{ kN} \quad \sigma^L = -100 \text{ MPa} \quad \sigma^U = 100 \text{ MPa} \quad l = 300 \text{ cm}$$

$$a = 0.666 \quad A_i^U = 30 \text{ cm}^2 \quad I_i = 3.2A_i^2 \quad Z_i = 1.452\, A^{1.5} \quad (i = 1,2)$$

The formulation of the two-objective optimization problem is

$$\text{Min } (V, \delta)$$

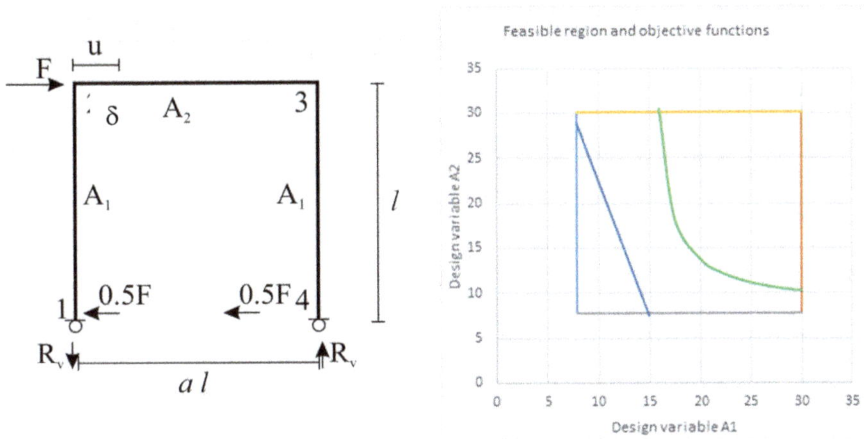

(a) (b)

Figure 4: Hinged portal frame. (a) Geometry and load; and (b) Feasible region and objective functions.

where

$$V = (2A_1 + aA_2)l \tag{15a}$$

$$\delta = \frac{(aI_1 + 2I_2)l^3 F}{12EI_1I_2} \tag{15b}$$

subject to

a) Vertical bars

$$\sigma = \frac{M}{Z_1} + \frac{N}{A_1} = \leq \sigma^U \tag{16a}$$

b) Horizontal bar

$$\sigma = \frac{M}{Z_2} + \frac{N}{A_2} = \leq \sigma^U \tag{16b}$$

$$A_i \leq A^U \qquad (i = 1, 2) \tag{16c}$$

where M, N are the bending moment and axial force at each point of the structure.

Fig. 4(b) shows the feasible region in the design variables domain and a realization of the volume V and the lateral displacement δ. In Fig. 5(a) a point appears where the gradient of both objective functions has the same direction. These points are on the line $A_2 = A_1$ and create the unconstrained Pareto points of segment A–B. The coordinates of point B are (7.9,

7.9). The rest of Pareto points are constrained and correspond to the short segment B–C located in the constraint $A_1 = 7.9\ cm^2$. The coordinates of point C are (7.9, 7.7).

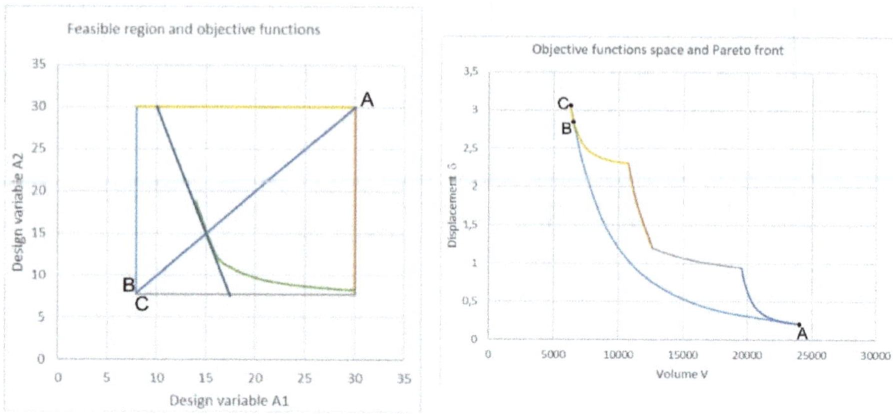

(a) (b)

Figure 5: (a) Pareto points; and (b) Pareto front.

The Pareto front in the objective functions space is presented in Fig. 5(b) and is the line A–B–C. The other lines correspond to values of the objective functions generated by the values of the design variables **X** on the boundaries of the feasible region in the design variables space. The application of the methodology presented allows obtaining the two classes of sensitivities of interest. Table 1 gives information of the numerical results, namely, the sensitivity of the design variables with regard to the objective functions at the Pareto points, the sensitivity of the objective function δ with regards to the volume V and also some approximate values of the objective functions and the design variables obtained with this approach and it can be seen that the level of accuracy is very high.

Table 1: Numerical results of the hinged portal frame.

A1	A2	dδ/dA1	dδ/dA2	dA1/dV	dA2/dV	dδ/dV	Disp. δ	Volume V	λ [0,1]	Pareto point	Disp δ αprox.	A1 aprox.	A2 aprox.
30	30	-0,010416667	-0,0034722	0,00125	0,00125	-1,736E-05	0,20833333	24000	0	Unconstrained			
22,7	22,7	-0,024044456	-0,0080148	0,00125	0,00125	-4,007E-05	0,36387277	18160	0,2	Unconstrained			
16,4	16,4	-0,064113058	-0,021371	0,00125	0,00125	-0,0001069	0,69968717	13096	0,4	Unconstrained			
12,5	12,5	-0,144346154	-0,0481154	0,00125	0,00125	-0,0002406	1,20192231	9992	0,6	Unconstrained	starting point		
10,8	10,8	-0,224510289	-0,0748368	0,00125	0,00125	-0,0003742	1,61348061	8624	0,7	Unconstrained			
9,01	9,01	-0,384519313	-0,1281731	0,00125	0,00125	-0,0006409	2,30967934	7208	0,8	Unconstrained			
11,9	11,9	-0,167954512	-0,0559848	0,00125	0,00125	-0,0002799	1,32963989	9500			1,324	11,85	11,85
11,3	11,3	-0,197530864	-0,0658436	0,00125	0,00125	-0,0003292	1,48148148	9000			1,444	11,23	11,23
7,9	7,9	-0,570441689	-0,1901472	0,00125	0,00125	-0,0009507	3,00432623	6320	0,87	Unconstrained			
7,9	7,9	-0,570441689	-0,1901472	0	0,005	-0,0009507	3,00432623	6320	0,87	Constrained A1=7,9			
7,9	7,7	-0,570441689	-0,2053521	0	0,005	-0,0010268	3,04385017	6280	1	Constrained A1=7,9			

4.2 Application example: Built-in portal frame

Another example of structural multi-objective optimization provides a different nature of the Pareto points in a structure with identical geometry. The built-in frame of Fig. 6(a) is subject to moments in both corners of the horizontal bar. The design variables, objective functions and constraints are the same as the previous example. The formulation of the two-objective optimization problem is as follows:

$E = 2 \cdot 10^5$ MPa $\quad M = 2$ kNm $\quad \sigma^U = 100$ MPa $\quad \sigma^L = -100$ MPa

$A_i^U = 15\ \text{cm}^2 \quad A_i^L = 2\ \text{cm}^2 \quad Z_i = 1.5 A_i^{1.5} \quad (i = 1,2) \quad l = 100$ cm $\quad a = 4$

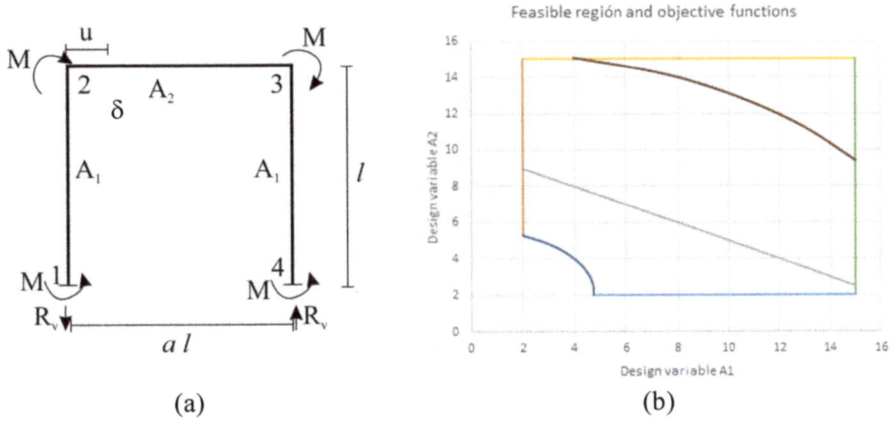

Figure 6: Built-in portal frame. (a) Geometry and loads; and (b) Feasible region and objective function.

The formulation of the two-objective optimization problem is

$$V = (2A_1 + aA_2)l \tag{17a}$$

$$\delta = \frac{aMl^2}{2E(aI_1 + 6I_2)} \tag{17b}$$

Subject to

$$\sigma = \frac{M}{Z_1} + \frac{N}{A_1} = \leq \sigma^U \tag{18a}$$

$$\sigma = \frac{M}{Z_2} + \frac{N}{A_2} = \leq \sigma^U \tag{18b}$$

$$A_i^L \leq A_i \leq A_i^U \qquad (i = 1, 2) \tag{18c}$$

The feasible region and in the design variables space and a realization of each objective function appear in Fig. 6(b). Fig. 7(a) shows a point in the feasible region at which the gradient of both objective functions has the same direction. These points are along the line $A_2 = 4A_1/3$ but do not produce Pareto points, they generate the yellow line D–E in the objective functions space presented in Fig. 7(b). It can be seen that the Pareto front corresponds to the points on the lines A–B and B–C. All are constrained Pareto points and are generated by the values of the design variables **X** on the constraints $A_1 = 15$ cm^2 and $A_2 = 2$ cm^2. Therefore, in this example all Pareto points are constrained.

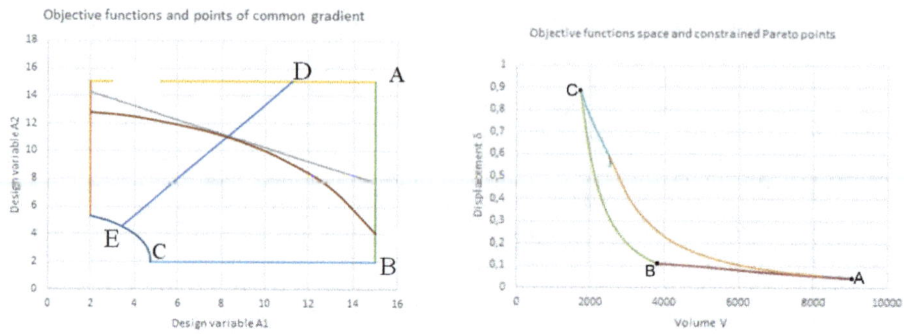

(a) (b)

Figure 7: (a) Pareto points; and (b) Pareto front.

The results produced by the methodology used in this research appear in Table 2.

Table 2: Numerical results of the built-in portal frame.

A1	A2	dδ/dA1	dδ/dA2	dA1/dV	dA2/dV	dδ/dV	Displa δ	Volume V	λ [0,1]		δ aprox	A1 aprox	A2 aprox
15	15	-0,00237037	-0,0035556	0	0,0025	-8,8889E-06	0,04444444	9000	0	Active constraint: A1=15			
15	10,57	-0,00487012	-0,0051453	0	0,0025	-1,2863E-05	0,06370582	7226	0,1	Active constraint: A1=15	start. Point		
15	3	-0,01318513	-0,0039555	0	0,0025	-9,8888E-06	0,1048218	4200		Active constraint: A1=15	0,102	15	3
15	2,9	-0,01328353	-0,0038522	0	0,0025	-9,6306E-06	0,10521221	4160		Active constraint: A1=15			
15	2,5	-0,01365333	-0,0034133	0	0,0025	-8,5333E-06	0,10666667	4000		Active constraint: A1=15			
15	2	-0,01405521	-0,002811	0	0,0025	-7,0276E-06	0,10822511	3800	0,2	Active constraint: A1=15	0,1077	15	2
15	2	-0,01405521	-0,002811	0,005	0	-7,0276E-05	0,10822511	3800	0,2	Active constraint: A2=2	start point		
12,6	2	-0,02315547	-0,0055088	0,005	0	-0,00011578	0,15150404	3322	0,5	Active constraint: A2=2	0,141	12,61	2
8,5	2	-0,06946032	-0,024522	0,005	0	-0,0003473	0,31964852	2499,54	0,75	Active constraint: A2=2			
4,72	2	-0,29512251	-0,1875779	0,005	0	-0,00147561	0,88406699	1744	1	Active constraint: A2=2			

Table 2 contains the same classes of numerical results presented in the other example. It can be observed than also in this case, with constrained Pareto points, the methodology performs appropriately and the approximate values of the objective functions and the design variables are very accurate.

5 CONCLUSIONS

The research presented confirms that, in multi-objective optimization, the Pareto front is composed of constrained Pareto points but also of unconstrained Pareto points with no active constraint. First order sensitivity analysis at the Pareto points gives information on the variations of the design variables with regard to the objective functions and sensitivity analysis of an objective function with regard to others at the Pareto points allows reducing the number of single optimization problems required to define the Pareto front and provides information about the relative influence of the objective functions. Two examples of structural multi-objective optimization with different classes of Pareto points are presented to demonstrate the efficiency of the methodology.

ACKNOWLEDGEMENTS

This research has received funding from the European Union´s Horizon 2020 research and innovation programme under the MORE&LESS project with grant agreement GA-

101006856. The funding received by the Galician Government through research grant ED131C 2021/33 is also acknowledged.

REFERENCES

[1] Osyczka, A., *Multicriterion Optimization in Engineering with FORTRAN Programs*, Ellis Horwood, 1984.

[2] Arora, J.S., *Introduction to Optimum Design*, Academic Press, 2016.

[3] Zhang, M.H., Domaszewski, M. & Fleury, C., An improved weighting method with multibounds formulation and convex programming for multicriteria structural optimization. *International Journal for Numerical Methods in Engineering*, **52**, pp. 889–902, 2001.

[4] Romero, C., Tamiz, M. & Jones, D.F., Goal and compromise programming and reference point methods formulations: Linkages and utility interpretations. *Journal of the Operations Research Society*, **49**(9), pp. 986–991, 1998.

[5] Ng, K.K.H., Lee, C.K.M., Chan, F.T.S. & Qin, Y., Robust aircraft sequencing and scheduling problem with arrival/departure delay using the min–max regret approach. *Transportation Research Part E: Logistics and Transportation Review*, **106**, pp. 115–136, 2017.

[6] Martínez, M.P., Messac, A. & Rais-Rohani, M., Manufacturability-based optimization of aircraft structures using physical programming. *AIAA Journal*, **39**(3), pp. 517–525, 2001.

[7] Cid Montoya, M., Costas, M., Díaz, J., Romera, L.E. & Hernández, S., A multi-objective reliability based optimization of the crashworthiness of a metallic-GFRP impact absorber using hybrid approximations. *Structural and Multidisciplinary Optimization*, **52**, pp. 827–843, 2015.

[8] Rodríguez-Segade, M., Hernández, S. & Díaz, J., Multi-level and multi-objective structural optimization for hypersonic vehicle design. *Aerospace Science and Technology*, **152**, 109346, 2024.

[9] Rodríguez-Segade, M., Steelant, J., Hernáández, S. & Díaz, J., Design optimization of multi-functional multi-lobe cryogenic fuel tank structures for hypersonic vehicles. *CEAS Space Journal*, **15**(6), pp. 813–826, 2023.

[10] Shau, T.N. & Chang, J.R., Multi-objective optimization of rotor-bearing system with critical speed constraints. *Journal of Engineering for Gas Turbines and Power*, **115**(2), pp. 246–255, 1993.

[11] Nieto, F., Cid Montoya, M. & Hernández, S., Shape optimization of tall buildings cross-section: Balancing profit and aeroelastic performance. *Structural Design of Tall and Special Buildings*, **31**(18), e1982, 2022.

[12] Koski, J., Multicriterion optimization in structural design. *New Directions in Optimum Structural Design*, eds E. Atrek, R.H. Gallagher, K.M. Ragsdell & O.C. Zienkiewicz, John Wiley, 1984.

SECTION 9
STRUCTURAL OPTIMISATION

DEALING WITH DISCRETE BINARY VARIABLES IN MINLP STRUCTURAL OPTIMIZATION

STOJAN KRAVANJA
Faculty of Civil Engineering, Transportation Engineering and Architecture, University of Maribor, Slovenia

ABSTRACT

This paper presents the rational use of discrete binary variables in mixed-integer non-linear programming (MINLP) structural optimization. MINLP is a powerful optimization method that integrates both discrete and continuous variables while considering non-linear constraints. MINLP is used in many areas, such as chemical engineering, energy systems, supply chain and logistics and financial portfolio management. MINLP also plays a crucial role in structural optimization. In this area, it can be used to optimize various structures such as steel, concrete, timber and composite systems such as frames, bridges and trusses. However, structural optimization problems using MINLP often involve an immense number of binary variables, sometimes in the tens of thousands, making the problems computationally intractable. To get a handle on this complexity, the number of binary variables must be reduced to a manageable size. This can be achieved by techniques such as incorporating special logical constraints, implementing multilevel MINLP strategies or performing preselection of binary variables. This paper specifically examines how many binary variables can be eliminated by pre-screening, a method that has been shown to be particularly effective. Five structural optimization case studies are examined: three large problems with up to tens of thousands of binary variables and two medium-sized problems with several hundred variables. The analysed structures include a steel truss, two hydraulic steel gates, a steel hall and a timber hall. The results show that pre-screening leads to a greater reduction in binary variables for large problems than for medium problems. On average, only about 5% of the original binary variables were retained in the MINLP optimization process after pre-screening. This significant reduction contributed to a significant improvement in the convergence speed of the OA/ER algorithm.
Keywords: structural optimization, mixed-integer non-linear programming, MINLP, discrete optimization, sizing optimization, topology optimization, multilevel strategy, pre-screening.

1 INTRODUCTION

Mixed-integer non-linear programming (MINLP) is a powerful optimization technique for solving combined discrete, non-linear and non-convex optimization problems. The latter can include both integer and continuous decision variables as well as non-linear objective functions and/or constraints. MINLP extends and combines the capabilities of integer programming (IP) or mixed-integer linear programming (MILP) with non-linear programming (NLP) by enabling more realistic modelling of complex systems where some decisions can be defined as binary or categorical. MINLP is therefore particularly valuable for real-world applications where decisions are not purely discrete or continuous, but are a combination of both, such as in structural optimization, industrial process optimization, supply chain management, and resource allocation.

MINLP is widely used in areas such as structural optimization, energy system design, chemical process optimization, logistics and finance, where decision making must consider both discrete decisions and non-linear relationships between variables. For example, in a manufacturing environment, a plant must decide which machines to turn on (discrete) and how much material to allocate to each process (continuous), while at the same time considering nonlinear physical or economic constraints.

In structural optimization, it is usually desirable to calculate for a given structure which structural elements make up the structure and where (discrete topology), which standard/

www.witpress.com, ISSN 1743-3509 (on-line)
doi:10.2495/HPSM250161

discrete profiles and strength classes/materials are optimal; and at the same time, it is also necessary to calculate some continuous dimensions (usually the lengths of some elements, e.g. diagonals). These design parameters are defined as variables that are calculated in the optimization when the objective function converges. The latter is usually defined as a mass or cost objective function. Binary 0–1 variables are defined for the discrete decisions. For example, a binary variable is assigned to each structural element, which is then selected to build the structure if the value of the calculated binary variable is 1, otherwise (0) it is removed.

For large MINLP problems, a major difficulty is the large number of discrete, binary 0–1 variables that are defined to solve discrete decisions. This number can be up to several tens of thousands, which usually prevents the problems from being solved successfully. To overcome this difficulty, the number of binary variables should be reduced. The latter can be done by defining special equations to reduce the number of binary variables, by using a suitable multilevel MINLP strategy and by preselecting the binary variables. The paper answers the question to what extent this reduction should be performed and how high the percentage of active binary variables is, that is particularly reduced by the pre-screening procedure.

2 APPLICATIONS OF THE MINLP

The MINLP method has become an indispensable tool for solving complex optimization problems in various industries, e.g. chemical engineering, energy system optimization, supply chain and logistics management, financial portfolio optimization and structural optimization.

In process optimization in chemical engineering, MINLP plays a crucial role in process design and optimization (see for example Grossmann and Kravanja [1] and Ye et al. [2]). These processes often involve discrete decisions – e.g. the selection of suitable equipment or the definition of operating modes – as well as continuous parameters such as flow rates, temperature and pressure that need to be fine-tuned. The aim is often to reduce energy consumption, increase product yield or reduce operating costs. Here, the models include non-linearities such as reaction kinetics and thermodynamic relationships, leading to more efficient and economically viable designs.

Energy production and distribution systems, ranging from electricity generation to gas distribution and renewable energy integration, represent comprehensive optimization problems (see for example Nguyen et al. [3] and Moretti et al. [4]). MINLP is often used to support both the planning and operational phases of these systems. Strategic decisions, e.g. where and whether to install energy-generating facilities such as power plants or wind turbines, are inherently discrete. Operational aspects, such as energy distribution and fuel consumption, on the other hand, involve continuous variables. These systems are further complicated by non-linear elements such as fuel cost curves and environmental compliance.

Supply chain optimization is another important application of MINLP that offers significant advantages in managing complex logistics networks (see for example Holler Branco et al. [5] and Billal et al. [6]). The goal is typically to reduce overall costs while ensuring customer demand and maintaining service standards across suppliers, warehouses and distribution hubs. Integer variables are used to model discrete decisions, such as the selection of warehouse locations, the choice of transportation methods and the number of vehicles. At the same time, continuous variables are used to capture aspects such as shipping volumes, production rates and inventory levels. Non-linearities often result from dynamic elements such as variable transportation costs, shifting demand patterns and lead times.

In the financial sector, MINLPs are often used for portfolio optimization to help investors maximize their returns under risk management constraints (see for example Díaz et al. [7] and Bayat et al. [8]). Integer variables are used to model decisions such as which assets to include in the portfolio, while continuous variables determine how much capital to allocate to each selected asset. The optimization process is often complicated by non-linear elements such as returns, individual risk preferences and transaction costs.

3 APPLICATIONS IN STRUCTURAL OPTIMIZATION

The MINLP method has become an indispensable tool for solving complex optimization problems for steel, concrete and timber structures such as frames, bridges, trusses, composite structures, energy efficient buildings and others.

MINLP optimization of frames includes discrete and non-linear continuous decisions such as the choice of material and cross-sectional profiles and the determination of load assignments. These optimizations are applied to both multi-storey frame structures and single-storey frames (halls). While the multi-storey frame structures are usually optimized for the fixed topology with a fixed number of storeys and bays, the optimization of hall structures is also performed to find an optimal number of portal frames, purlins and rails (topology) (see Klanšek et. al. [9] and Kravanja et. al. [10]).

MINLP optimization of bridges requires the simultaneous calculation of discrete design variables (e.g. beam types, column sizes) and continuous parameters (e.g. load conditions, displacement limits). MINLP has been applied to bridge design problems, where it optimizes both structural configuration and material selection considering traffic and wind loads as well as safety requirements (see Kusano et al. [11] and Jin et al. [12]).

The optimization of trusses typically requires a combination of discrete and non-linear continuous decisions – for example, the optimization of the number of members (topology), the selection of materials or cross-sectional profiles, and the determination of load distribution (see for example Ohsaki and Katoh [13] and Šilih et al. [14]). MINLP has been widely used in this field to find optimal configurations that reduce weight or minimize cost while meeting structural safety and performance requirements. Its ability to deal with both variable types and non-linear constraints makes it particularly well suited to handle the complexity of truss optimization.

MINLP optimization of composite structures combines both continuous variables (such as intermediate spacing between I-beams) and integer variables (such as the selection of discrete materials and standard sizes) while dealing with nonlinear relationships (such as resistance conditions and deflection constraints). It is useful for finding the best design that meets performance requirements while minimizing weight, cost, or other objectives (see Kravanja et al. [15] and Kırdar et al. [16]).

MINLP optimization of energy efficient buildings balances structural integrity, thermal performance and energy consumption. The MINLP is used to optimize the structural and energy systems of buildings to ensure minimum energy consumption while maintaining structural strength and stability (see for example Lešnik Nedelko et al. [17] and Kravanja et al. [18]).

MINLP optimization is also performed for the optimization of structures for seismic design, for offshore structures, hydrotechnical structures (steel gates and high-pressure penstocks), cranes and others.

4 SOLVING THE MINLP PROBLEM

MINLP optimization problems are combined discrete, continuous, non-linear and non-convex problems that can be mathematically formulated as follows:

$$\min f(\boldsymbol{x},\boldsymbol{y})$$

Subject to:

$$\boldsymbol{h}(\boldsymbol{x},\boldsymbol{y}) = \mathbf{0} \qquad \text{(MINLP)}$$

$$\boldsymbol{g}(\boldsymbol{x},\boldsymbol{y}) \leq \mathbf{0}$$

$$\boldsymbol{x} \in X = \{\boldsymbol{x} \mid \boldsymbol{x} \in R^{n}; \boldsymbol{x}^{lo} \leq \boldsymbol{x} \leq \boldsymbol{x}^{up}\}$$

$$\boldsymbol{y} \in Y = \{0, 1\}^{m}$$

where:

- $\boldsymbol{x}$ is a vector of continuous variables,
- $\boldsymbol{y}$ is a vector of discrete/binary variables,
- $f(\boldsymbol{x},\boldsymbol{y})$ is (usually) a non-linear objective function,
- $\boldsymbol{h}(\boldsymbol{x},\boldsymbol{y}) = \mathbf{0}$ is a set of equality constraints, and
- $\boldsymbol{g}(\boldsymbol{x},\boldsymbol{y}) \leq \mathbf{0}$ is a set of inequality constraints.

Note that at least one function (or objective or constraint) must be non-linear. The presence of both discrete and continuous variables significantly increases the difficulty of the problem, since non-linear programming (NLP) solvers cannot explicitly handle discrete variables, and integer programming (IP) solvers are not designed for non-linear constraints. These problems can be effectively solved with the following MINLP algorithms:

- Generalized Benders Decomposition by Benders [19] and Geoffrion [20],
- Outer-Approximation/Equality-Relaxation (OA/ER) algorithm by Kocis and Grossmann [21],
- Sequential Linear Discrete Programming method by Olsen and Vanderplaats [22] and Bremicker et al. [23],
- Extended Cutting Plane algorithm by Westerlund and Petterson [24], and
- the LP/NLP Branch and Bound algorithm by Quesada and Grossmann [25].

4.1 The modified OA/ER algorithm

To overcome the challenges posed by non-convexities in MINLP problems, the improved version, the modified outer-approximation/equality-relaxation (OA/ER) algorithm developed by Kravanja and Grossmann [26], is used together with the MIPSYN computational platform [27]. This algorithm builds on the basic OA/ER method by introducing advanced convexification strategies to mitigate the negative effects of non-convexities during the iterative optimization process. This process alternates between two types of optimization methods: nonlinear programming (NLP) sub-problems and mixed-integer linear programming (MILP) main problems (see Fig. 1).

In this optimization process, each NLP sub-problem focuses on optimizing the continuous variables associated with a mechanical structure whose topology, material selection, and standard dimensional parameters were specified in the previous MILP step. These NLP steps contribute to the refinement of the global approximation of the entire mechanical design space. After solving each NLP, the MILP main problem takes over. It performs a discrete optimization of the updated linear model of the superstructure to propose a new configuration – topology, materials and standard dimensions – for the next NLP iteration. In this two-layer

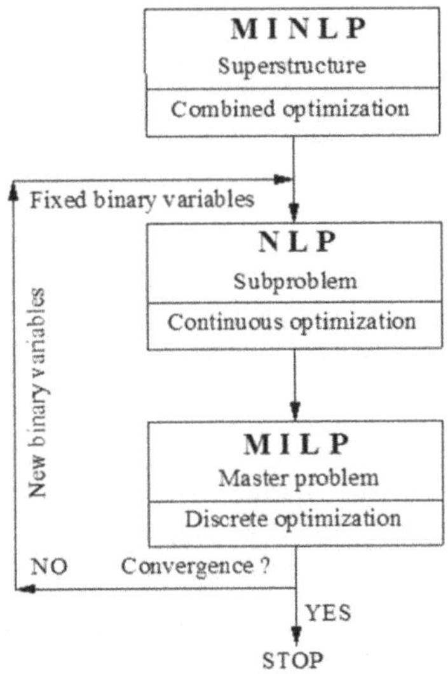

Figure 1: Steps of the OA/ER algorithm.

process, the main MILP problem provides a lower bound, while the NLP sub-problem provides an upper bound on the overall MINLP objective. Convergence is achieved when the two bounds coincide, signifying the optimal solution for convex problems. In non-convex cases, optimization is complete when successive NLP iterations no longer yield any improvements.

The NLP steps are solved using the GAMS/CONOPT solver, which applies the generalized reduced gradient algorithm [28]. Meanwhile, the MILP problems are treated with GAMS/CPLEX, which uses a branch and bound method [29]. GAMS (general algebraic modelling system) is proposed for modelling [30].

5 HANDLING OF DISCRETE BINARY VARIABLES

In discrete optimization, a major problem is the large number of discrete binary 0–1 variables defined in the problems, since the number of all combinations (e.g. different designs due to different topologies, materials, and standard dimensions) is 2^m, where m is the number of binary variables in the problem. This problem is even more pronounced with MINLP, which contains both linear and nonlinear functions. In order to obtain a better result or an optimal solution for large problems, it is necessary to reduce the binary variables. This is done by introducing special logical equations to reduce the number of binary variables, by using a suitable MINLP strategy and by preselecting the binary variables.

5.1 Special logical equation for reducing the binary variables

A special logical equation in which binary variables are multiplied and summed with different integer scalars can reduce the number of binary variables in problems. With eqn (1), for example, it is possible to define eight possible integers of NUM with a choice of only three binary variables:

$$NUM = y_1 + 2y_2 + 4y_3 \tag{1}$$

If all three calculated binary 0–1 variables are equal to 0, the number *NUM* is also equal to zero according to eqn (1). *NUM* is 1 if y_1 is 1 and the other two binary variables are 0. *NUM* is 2 if y_2 is 1 and the other two are 0. *NUM* is 3 if y_1 and y_2 are 1 and y_3 is 0. *NUM* is 4 if y_3 is 1 and the other two variables are 0. *NUM* is 5 if y_1 and y_3 are 1 and y_2 is 0. *NUM* is 6 if y_2 and y_3 are 1 and y_1 is 0. And finally, *NUM* is 7 if all binary variables are calculated as 1. In this way, eqn (1) requires fewer binary variables for the same condition than eqn (2):

$$NUM = y_1 + y_2 + y_3 + y_4 + y_5 + y_6 + y_7 \tag{2}$$

With a very high number of *NUM*, see eqn (3), this notation allows the use of significantly fewer binary variables, resulting in a much faster calculation.

$$NUM = \sum_i^N 2^{(i-1)} \cdot y_i \quad i \in I \tag{3}$$

5.2 Linked multilevel hierarchical strategy (LMHS)

To streamline the solution process and support faster convergence within the OA/ER framework, the multilevel MINLP strategies, such as the linked multilevel hierarchical strategy (LMHS) [31], [32], are used.

LMHS decomposes the original MINLP problem into a structured hierarchy of levels, both in terms of variable space and optimization stages. Instead of dealing with the entire set of discrete variables Y_{TOT} at once, the discrete decision space is expanded step by step. At each level, a new subset of binary variables is introduced and pre-evaluated, while variables from other subsets are approximated by relaxed 0–1 representations. With each level, the accumulated outer approximations refine the model and help generate stronger lower bounds for the next stage of optimization.

In structural optimization, this hierarchy is generally composed of four decision levels:

- Level 1 focuses on discrete choices concerning the topology of the structure; here only binary variables y from the set for topology determination Y_{top} are used, while all other design parameters (materials, discrete dimensions) are continuous: $y \in Y_{top}$.
- Level 2 handles discrete decisions related to material qualities; here discrete optimization of topology and material is performed simultaneously, while all other parameters (discrete dimensions) remain continuous; binary variables are enriched here with the set of binary variables Y_{mat} for discrete materials: $y \in (Y_{top} \cup Y_{mat})$.
- Level 3 adds discrete decisions for standard dimensions; the simultaneous discrete optimization of topology, material and standard dimensions is performed at this level; a set of binary variables for standard dimensions Y_{st} is added: $y \in (Y_{top} \cup Y_{mat} \cup Y_{st})$.
- Level 4 refines the design using rounded continuous dimension values; the entire discrete topology, material, standard dimension and rounded dimension optimization is then applied; since a set of binary variables for rounded dimensions Y_{rd} is added, the entire set of binary variables Y_{TOT} is used here: $y \in (Y_{TOT} = Y_{top} \cup Y_{mat} \cup Y_{st} \cup Y_{rd})$.

The optimization process starts with the level 1 decisions (topology optimization), assuming relaxed values for the materials and dimensions. Once the optimal selection is made, the algorithm moves to the next level (level 2) to optimize both the discrete topology and the material selection. Once the optimal solution is found, the optimization process moves to discrete optimization at level 3 (topology, materials and standard dimensions). At the final level (level 4), the model incorporates all discrete variables Y_{TOT}, including the rounded dimensional alternatives derived from the earlier optimization phases, and the entire discrete optimization is performed.

Like the OA/ER algorithm, the LMHS ensures that the final solutions for convex problems are globally optimal (for non-convex problems they are usually local optima and practically acceptable).

5.3 Pre-screening of binary variables

Since the total number of binary variables, especially those linked to rounded dimensions, can run into tens of thousands, LMHS uses an automatic variable reduction procedure. This technique filters out less relevant binary variables and focuses only on those that closely match the previously obtained continuous solutions. For example, for each design parameter (topology, material or dimension), 3 (or 2) discrete variables below and 3 (or 2) above the calculated continuous value are considered. In this way, only 6 (or 4) binary variables are used to calculate a design parameter. This pre-screening of binary variables can be performed at any level and for any design parameter.

6 ANALYSIS OF PRE-SCREENED BINARY VARIABLES IN EXAMPLES

How many binary variables are reduced by pre-screening varies from problem to problem. It is assumed that the percentages of reduced-active binary variables are low for large optimization problems and higher for medium or small problems. To determine the number and percentage of reduced-active binary variables in structural optimization problems, some previously performed and published MINLP examples are analysed [32]–[34].

For example, Kravanja [32] presents the simultaneous MINLP optimization of mass, shape and standard dimensions of a simply supported 45-bar steel truss over a span of 30 m (see Fig. 2). Since the topology of the structure was not changed in the optimization, a single level of LMHS strategy was applied (with all binary variables organised in a single set). In this problem, 2,093 binary variables were reduced to 160 by the pre-screening procedure. An optimal mass of 2,250 kg was determined.

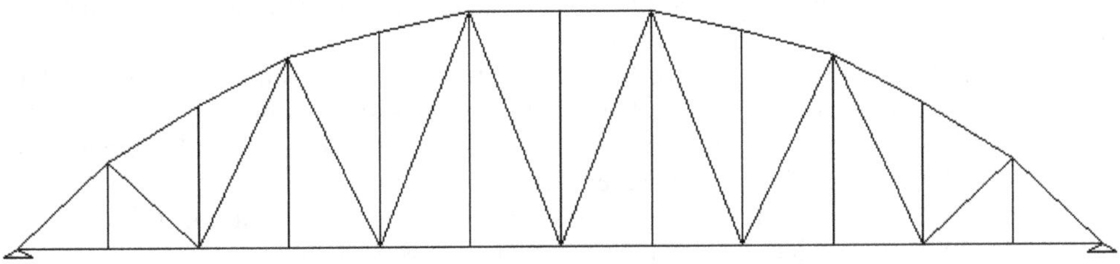

Figure 2: Optimal shape of the truss [32].

In Kravanja [32], the simultaneous MINLP optimization of costs, topology, standard dimensions and rounded dimensions of a roller hydraulic steel gate is presented. The gate was manufactured for the Hydroelectric Project Blanda in Iceland (see Fig. 3). The optimization problem was very extensive as it involved 19,622 binary and 59,097 continuous variables. A three-level LMHS strategy was applied (level 1: topology, level 2: standard dimensions, level 3: rounded dimensions) for the steel S275 (St 44-3) used. 19,622 of the binary variables were pre-screened to 224 variables. The optimal self-manufacturing costs of €8,804 were obtained.

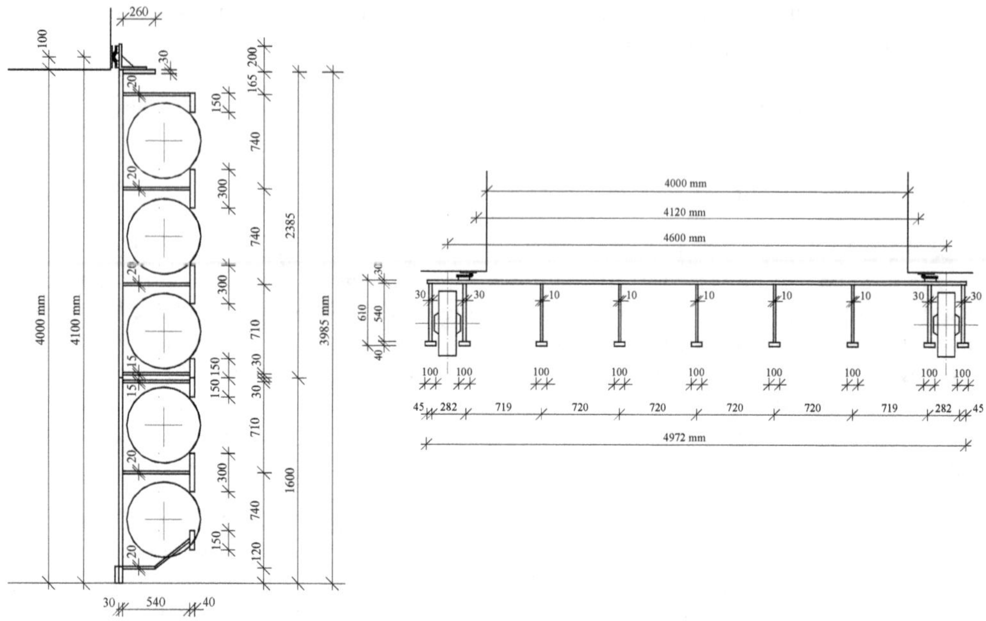

Figure 3: Vertical and horizontal cross-sections of the gate [32].

The MINLP optimization of another roller hydraulic steel gate is reported: the optimization of production costs, topology, material, standard and rounded dimensions of the Intake gate. Eight such gates were manufactured for Aswan II, Egypt [33]. A four-level LMHS strategy was used (level 1: topology, level 2: materials, level 3: standard dimensions, level 4: rounded dimensions). 35,052 of the binary variables were pre-screened to 666 variables which enabled the calculation of the optimal result. The optimal self-manufacturing costs of $48,811 were determined in this way.

Kravanja and Žula [34] present the MINLP optimization of production costs, topology, material, standard dimensions and rounded dimensions for a single-storey steel hall structure with a span of 16 m, a length of 80 m and a height of 4.5 m. A two-level LMHS strategy was applied (level 1: topology plus materials, level 2: standard dimensions together with rounded dimensions). 360 binary variables were pre-screened/reduced to 27 variables. The optimal result for the minimal production costs of the structure was €100,245 (see Fig. 4).

In Kravanja and Žula [34], the simultaneous MINLP optimization of material costs, topology, materials, standard dimensions and rounded dimensions of a timber hall structure with the same global geometry as above is also presented. A two-level LMHS strategy was applied (the same as above). Here, 426 binary variables were pre-screened to 33 variables. The minimal material cost of the hall structure reached €138,207 (see Fig. 5).

While the first three examples (the steel truss and the gates) represent large and very extensive MINLP optimization problems with several thousand discrete binary variables, the last two examples represent medium-sized problems. The percentages of binary variables reduced by the pre-screening procedure are listed for each example in Table 1. As hypothesized, the percentage of reduced-active binary variables is low for the large problems: 3.5%, while it is higher for the medium problems: 7.6%. The average percentage for all

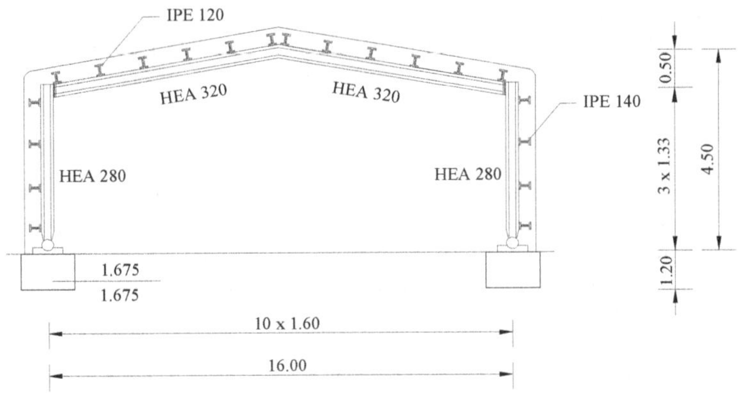

Figure 4: Optimal structure of the steel hall frame [34].

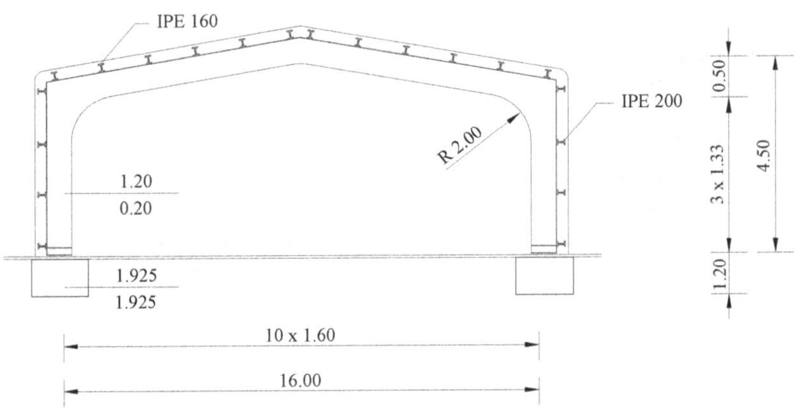

Figure 5: Optimal structure of the timber hall frame [34].

Table 1: Percentage of reduced-active binary variables by the pre-screening procedure.

Example: Size of the problem	Type of structure and reference	No. of LMHS levels	Total no. of binary variables	No. of prescreened binary variables	Percentage of reduced-active binary variables
1: large	Truss [32]	1 level	2,093	160	7.6%
2: large	Gate Blanda [32]	3 levels	19,622	224	1.1%
3: large	Gate Aswan [33]	4 levels	35,052	666	1.9%
Average percentage of reduced binary variables for large problems:					**3.5%**
4: medium	Steel hall [34]	2 levels	360	27	7.5%
5: medium	Timber hall [34]	2 levels	426	33	7.7%
Average percentage of reduced binary variables for medium problems:					**7.6%**
Average percentage of reduced binary variables for all problems:					**5.2%**

structures/examples considered is 5.2%. This means that not the entire sets of binary variables, but only about 5% of them were involved in the MINLP optimization processes after pre-screening, which significantly accelerated the convergences of the OA/ER algorithm.

7 CONCLUSIONS

This paper deals with MINLP, which is used in structural optimization. MINLP is a combined discrete and continuous optimization technique which also takes into account nonlinear constraints. For this reason, both discrete and continuous variables are used. MINLP is used to solve optimization problems in chemical engineering, the optimization of energy systems, supply chain and logistics management, the optimization of financial portfolios and also in structural optimization. In structural optimization, MINLP can be used for the optimization of steel, concrete, timber and composite structures such as trusses, floor systems, frames, bridges and others.

MINLP problems in structural optimization can contain up to several tens of thousands of binary variables, which makes the problems unsolvable. For this reason, the number of binary variables should be reduced to a reasonable level by introducing special logical equations, or by using a suitable multilevel MINLP strategy or by pre-screening the binary variables. The article answers the question of how many binary variables are reduced by pre-screening. Five MINLP optimization problems are analysed, three large problems with up to several tens of thousands of binary variables and two medium problems with several hundred variables. A steel truss, two hydraulic steel gates, a steel hall structure and a timber hall structure are analysed. It was found that after pre-screening, large problems have a lower percentage of active binary variables than medium problems. It was also found that only about 5% of the binary variables were involved in the MINLP optimization processes after pre-screening, which significantly accelerated the convergences of the OA/ER algorithm.

ACKNOWLEDGEMENT

The author is grateful for the support of the Slovenian Research and Innovation Agency (Program P2-0129).

REFERENCES

[1] Grossmann, I.E. & Kravanja, Z., Mixed-integer nonlinear programming techniques for process systems engineering. *Computers and Chemical Engineering*, **19**(s1), pp. 189–204, 1995.

[2] Ye, Y., Grossmann, I.E. & Pinto, J.M., Mixed-integer nonlinear programming models for optimal design of reliable chemical plants. *Computers and Chemical Engineering*, **116**, pp. 3–16, 2018.

[3] Nguyen, K.D., Dai, P.D., Vu, D.Q., Cuong, B.M., Tuyen, V.P. & Li, P., A MINLP model for optimal localization of pumps as turbines in water distribution systems considering power generation constraints. *Water*, **12**(7), 12071979, 2020.

[4] Moretti, L., Manzolini, G. & Martelli, E., MILP and MINLP models for the optimal scheduling of multi-energy systems accounting for delivery temperature of units, topology and non-isothermal mixing. *Applied Thermal Engineering*, **184**, 116161, 2021.

[5] Holler Branco, J.E., Holler Branco, D., de Aguiar, E.M., Filho, J.V.C. & Rodrigues, L., Study of optimal locations for new sugarcane mills in Brazil: Application of a MINLP network equilibrium model. *Biomass and Bioenergy*, **127**, 105249, 2019.

[6] Billal, M.M., Sebastian, R.M. & Kumar, A., Developing a GIS-based MINLP framework for optimizing the waste-to-energy supply chain with plant scale considerations. *Journal of Cleaner Production*, **490**, 144709, 2025.

[7] Díaz, J., Cortés, M., Hernández, J., Clavijo, Ó., Ardila, C. & Cabrales, S., Index fund optimization using a hybrid model: Genetic algorithm and mixed-integer nonlinear programming. *The Engineering Economist*, **64**(3), pp. 298–309, 2019.

[8] Bayat, M., Hooshmand, F. & MirHassani, S.A., Drawdown minimization in asset portfolio selection: MINLP models and efficient cross-entropy algorithm. *Intl. Trans. in Op. Res.*, **0**, pp. 1–25, 2024.

[9] Klanšek, U., Žula, T., Kravanja, Z. & Kravanja, S., MINLP optimization of steel frames. *Advanced Steel Construction*, **3**(3), pp. 689–705, 2007.

[10] Kravanja, S., Turkalj, G., Šilih, S. & Žula, T., Optimal design of single-story steel building structures based on parametric MINLP optimization. *Journal of Constructional Steel Research*, **81**, pp. 86–103, 2013.

[11] Kusano, I., Montoya, M.C., Baldomir, A., Nieto, F., Jurado, J.A. & Hernández, S., Reliability based design optimization for bridge girder shape and plate thicknesses of long-span suspension bridges considering aeroelastic constraint. *Journal of Wind Engineering and Industrial Aerodynamics*, **202**, 104176, 2020.

[12] Jin, S.S., Park, Y.S., Kim, S.T. & Park, Y.H., Model updating based on mixed-integer nonlinear programming under model-form uncertainty in finite element model. *Engineering with Computers*, **37**, pp. 3699–3725, 2021.

[13] Ohsaki, M. &·Katoh, N., Topology optimization of trusses with stress and local constraints on nodal stability and member intersection. *Struct. Multidisc. Optim.*, **29**, pp. 190–197, 2005.

[14] Šilih, S., Kravanja, Z. & Kravanja, S., The MINLP approach to topology, shape and discrete sizing optimization of trusses. *Applied Sciences*, **12**(3), p. 36, 2022.

[15] Kravanja, S., Žula, T. & Klanšek, U., Multi-parametric MINLP optimization study of a composite I beam floor system. *Engineering Structures*, **130**, pp. 316–335, 2017.

[16] Kırdar, G., Ozturk, D.T. & Testik, M.C., Regression-based multiobjective optimization approaches for the autoclave composite parts placement problem: A mixed integer nonlinear programming model and an evolutionary algorithm. *Soft Computing*, **27**, pp. 14045–14062, 2023.

[17] Lešnik Nedelko, M., Kravanja, S., Premrov, M. & Žegarac Leskovar, V., Optimal design of timber-glass upgrade modules for vertical building extension from the viewpoints of energy efficiency and visual comfort. *Applied Energy*, **270**, pp. 1–15, 2020.

[18] Kravanja, S., Klanšek, U. & Žula, T., Mass, direct cost and energy life-cycle cost optimization of steel-concrete composite floor structures. *Applied Sciences*, **11**(21), pp. 1–23, 2021.

[19] Benders, J.F., Partitioning procedures for solving mixed-variables programming problems. *Numerische Mathematik*, **4**, pp. 238–252, 1962.

[20] Geoffrion, A.M., Generalized Benders decomposition. *Journal of Optimization Theory and Applications*, **10**(4), pp. 237–260, 1972.

[21] Kocis, G.R. & Grossmann, I.E., Relaxation strategy for the structural optimization of process flowsheets. *Ind. Engng Chem. Res.*, **26**, pp. 1869–1880, 1987.

[22] Olsen, G.R. & Vanderplaats, G.N., Method for nonlinear optimization with discrete design variables. *AIAA Journal*, **27**(11), pp. 1584–1589, 1989.

[23] Bremicker, M., Papalambros, P.Y. & Loh, H.T., Solution of mixed-discrete structural optimization problems with a new sequental linearization method. *Computers and Structures*, **37**(4), pp. 451–461, 1990.
[24] Westerlund, T. & Petterson, F., An extended cutting plane method for solving convex MINLP problems. *Computers Chem. Eng.*, **19**, pp. 131–136, 1995.
[25] Quesada, I. & Grossmann, I.E., An LP/NLP based branch and bound algorithm for convex MINLP optimization problems. *Computers Chem. Engng*, **16**, pp. 937–947, 1992.
[26] Kravanja Z. & Grossmann I.E., New developments and capabilities in PROSYN: An automated topology and parameter process synthesizer. *Computers Chem. Eng.*, **18**, pp. 1097–1114, 1994.
[27] Kravanja, Z., Challenges in sustainable integrated process synthesis and the capabilities of an MINLP process synthesizer MipSyn. *Comput. Chem. Eng.*, **34**(11), pp. 1831–1848, 2010.
[28] Drudd, A.S., CONOPT: A large-scale GRG code. *ORSA J. Comput.*, **6**, pp. 207–216, 1994.
[29] CPLEX User Notes, ILOG inc.
[30] Brooke, A., Kendrick, D. & Meeraus, A., *GAMS: A User's Guide*, Scientific Press: Redwood City, CA, 1988.
[31] Kravanja, S, Soršak, A. & Kravanja, Z., Efficient multilevel MINLP strategies for solving large combinatorial problems in engineering. *OPTimization and Engineering*, **4**(1/2), pp. 97–151, 2003.
[32] Kravanja, S., MINLP optimization approach. *International Conference on Computer Aided Optimum Design of Structures*, pp. 15–24, 2003.
[33] Kravanja, S., Šilih, S. & Kravanja, Z., The multilevel MINLP optimization approach to structural synthesis: The simultaneous topology, material, standard and rounded dimension optimization. *Advances in Engineering Software*, **36**(9), pp. 568–583, 2005.
[34] Kravanja, S. & Žula, T., Optimization of steel and timber hall structures. *International Conference on High Performance and Optimum Structures and Materials Encompassing Shock and Impact Loading, HPSM/OPTI/SUSI 2022*, pp. 35–46, 2022.

OPTIMUM DESIGN OF COMPOSITE I-BEAM WING SPARS WITH A CORRUGATED WEB

DOGAN COK & FAZIL O. SONMEZ
Department of Mechanical Engineering, Bogazici University, Istanbul, Turkey

ABSTRACT

This study develops a methodology for optimally designing composite I-beam wing spars with a wavy corrugated web under aerodynamic loads to minimize mass. Using the Hurkus Advanced Training Plane as a reference, the lift forces calculated from aircraft mass are applied directly to the spar as a transverse load. IM7/8552 carbon-fibre reinforced composite is selected for its high strength-to-weight ratio. Given that web buckling under shear is the critical failure mode, a buckling analysis determines the failure load. Codes are developed using ANSYS Parametric Design Language to implement the optimization algorithm and carry out structural analyses to determine the load capacity of the spar. The web is modelled as a multi-period wave defined by four parameters (initial wavelength, wavelength increment rate, initial amplitude and final amplitude). The objective is to minimize spar mass subject to buckling constraint. The modified simulated annealing algorithm – a stochastic global search method – identifies the optimal web shape. Results show a significant mass reduction while maintaining adequate strength compared to conventional spars.

Keywords: wing spar design, fibre composites, nonlinear buckling, structural optimization, finite element analysis (FEA).

1 INTRODUCTION

Aircraft wings, which house fuel tanks and carry significant loads during all flight phases, rely on structural components like ribs and spars (see Fig. 1). Spars, attached to the fuselage and extending across the wing, primarily endure upward bending from lift and downward bending due to the wing's weight [1]. Their performance is crucial for overall aircraft safety and efficiency.

Spars vary in shape and material based on specific needs. Small aircraft may use wood for its cost-effectiveness and light weight, while metals such as aluminium alloys or titanium are chosen for higher strength. However, carbon-fibre-reinforced polymers are popular in modern aircraft due to their superior strength-to-weight ratios.

1.1 Literature review

Prior studies explored various spar configurations, single spars [3], two spars per wing [4], [5] and multi spars [6], I-shapes with flat web [1], [3]–[7], C-shapes [1], [6] and bar shapes [7]. Different materials were tried such as wood [7], cobalt, steel [4], aluminium alloy [4], [5] and composite [1], [3]. Bhachu et al. [6] optimized aluminium spar tolerances for cost and fatigue resistance. Visnjic et al. [3] found thicker composite webs reduce shear lag but increase weight. Ajith et al. [5] highlighted front spar placement's impact on aeroelastic efficiency in high-aspect-ratio wings.

1.2 Problem statement and objective

The I-beam is one of the most commonly used types of spars in aircraft wings. Lift forces acting on the wing generate significant normal stresses in the flanges, while the web primarily

www.witpress.com, ISSN 1743-3509 (on-line)
doi:10.2495/HPSM250171

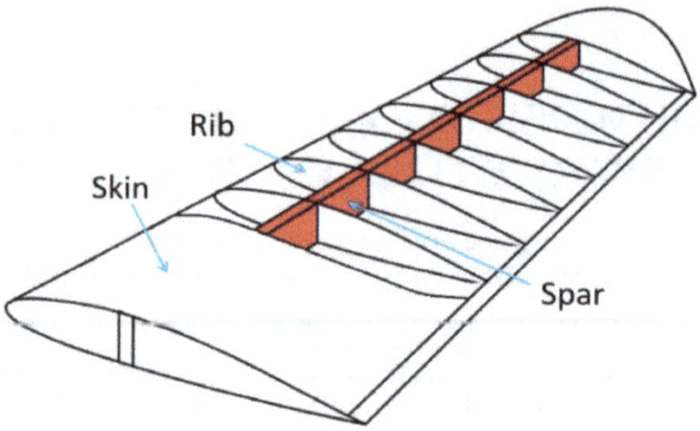

Figure 1: Schematic representation of aircraft wing spars [2].

endures transverse shear stresses. Although continuous fibre-reinforced composites provide high load-carrying capacity with thin laminates, a slender web can buckle under high transverse loads due to diagonal compressive stresses. Traditional solutions – such as adding stiffeners or increasing web thickness – add undesirable weight.

Alternatively, designing the web as a corrugated structure (see Fig. 2) offers improved buckling resistance. This study aims to establish an optimal design methodology for wing spars with wavy corrugated webs for minimum mass. Although manufacturing costs might rise, these are offset by fuel savings. Conventional constant corrugation is inefficient because high amplitudes and dense waves are only needed near the fuselage. The proposed optimization method allows variable corrugation along the spar, eliminating unnecessary mass associated with dense corrugation away from the fuselage.

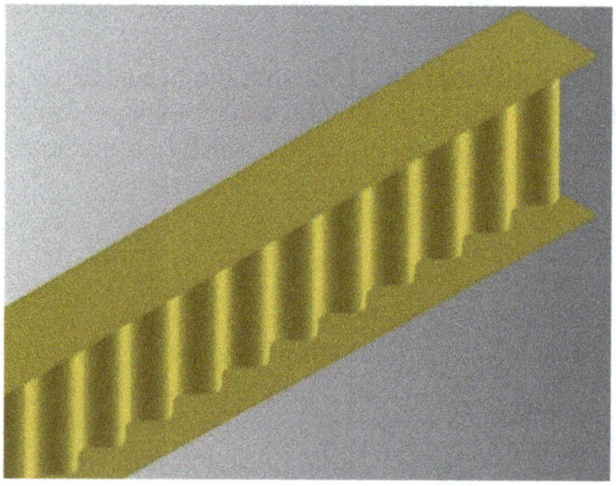

Figure 2: Isometric view of the I-beam wing spar model with a wavy corrugated web.

2 METHODOLOGY

A component-based design optimization approach is adopted to manage the complexity of the problem and reduce computational cost. In this approach, individual structural parts of the aircraft (e.g., spars, ribs) are optimized separately via a global search algorithm, considering many design variables and allowing large changes in their values. The effects of the adjacent parts are accounted for by imposing appropriate boundary conditions. In this way, computational times for design optimization will not be prohibitively long. Once individual components are optimized, the whole wing containing all the structural parts, e.g., spars, ribs, skin, ailerons, flaps, is optimized starting from the optimized shapes of the individual parts, but this time, the ranges of values that can be assigned to the variables will be small and some of the parameters may be taken as constant. The loading on the full wing model should reflect the loading conditions in the real wing structural tests; thus, the interactions between the individual parts can be more realistically simulated.

This study focuses solely on the modelling and design optimization of an I-beam spar beam, more specifically, its web. Thus, merely the spar beam is modelled rather than the entire airplane or wing. The loads transmitted by the other parts are taken into account by applying appropriate boundary conditions to the spar model.

The spar beam is modelled as a cantilevered shell beam with a wavy corrugated web subjected to a nonuniformly distributed lift load. Considering Hurkus Advanced Training Plane, the load on the spar and its dimensions are defined. The geometric design is parameterized to easily adapt to other planes. A finite element model of the spar is developed with mesh sizes determined through a convergence study.

The design optimization proceeds in two steps. First, the web shape is optimized to minimize weight while ensuring buckling resistance. The optimization variables are chosen to be the parameters defining the web shape while the fibre orientations are kept constant. In the second step, fibre orientations are optimized to maximize buckling strength, with the shape fixed. To avoid local optima inherent in many structural optimization problems, the modified simulated annealing algorithm [8], [9] is employed for a global search of the optimal values of design variables.

3 DESIGN PARAMETERS OF THE SPAR BEAM

A baseline spar model is established for web design optimization. The primary dimensions and loads are determined from the Hurkus Advanced Training Aircraft, with data provided by Turkish Aerospace Industries (TAI). The tapered I-beam spar has a length L of 4.0 m (each spar, considering the 10.91 m wingspan and fuselage width), a root height h_r of 200 mm, a tip height h_t of 150 mm, a root flange width b_r of 100 mm, a tip flange width b_t of 75 mm and a tip height t_h of 25 mm, as illustrated in Fig. 3.

The corrugated web is the most complex part, while the straight flanges can be designed conventionally or separately optimized. Hence, the geometric design of the flange is determined through a parametric study, while the shape and fibre orientations of the web are optimized by developing an optimization algorithm. In order for the laminate to have sufficient strength against the normal stresses, the fibres should mainly be oriented along the length of the beam. Accordingly, the layup sequence of the flanges is chosen as $[45/0/{-45}/0]13_s$. In the web, initial fibre angles of ±45° are assumed to counteract transverse shear and diagonal compressive stresses; these angles are later optimized in the second stage.

3.1 Design parameters of the web

The corrugated web is defined by a wave-like spline curve with variable wavelength and amplitude. Four parameters characterize its shape: the initial wavelength w_1, wavelength increment rate i, initial amplitude a_1 and final amplitude a_f (see Fig. 4). The spar acts as a cantilevered beam under a transverse distributed load, with maximum shear force at the fixed

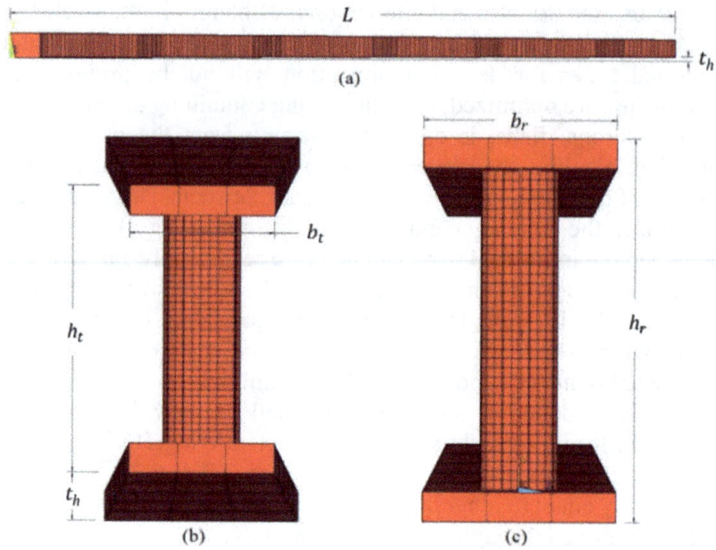

Figure 3: Wing spar dimensions. (a) Side view; (b) Front view; and (c) Rear view.

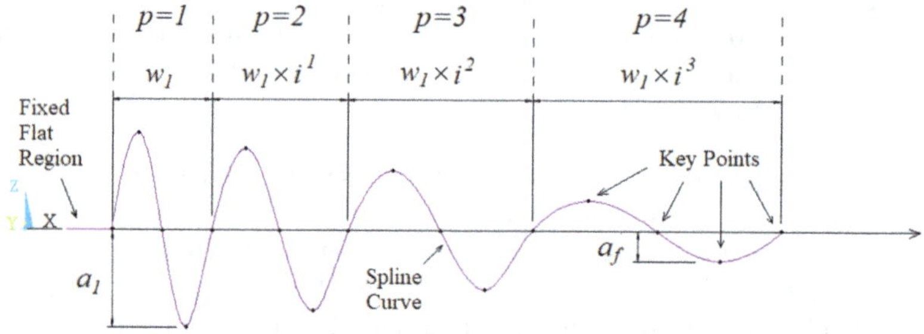

Figure 4: The parameters defining the shape of the corrugated web shape.

root. Therefore, a higher amplitude and shorter wavelength near the root provide greater buckling resistance, with the waveform gradually transitioning to longer wavelengths and lower amplitudes toward the free end.

4 FINITE ELEMENT MODELING

A finite element model of the spar beam is developed in ANSYS to simulate its mechanical behaviour and calculate the buckling load.

4.1 Element type and mesh

The spar, modelled as a thin-walled shell rather than a 3D solid for computational efficiency, meshed with SHELL181 four-node elements. The model has three main components: top flange, bottom flange, and web. Each component is meshed separately, and bonded contacts between flanges and web edges are defined using CONTA175 and TARGE170 elements with node-to-surface contact (Fig. 5).

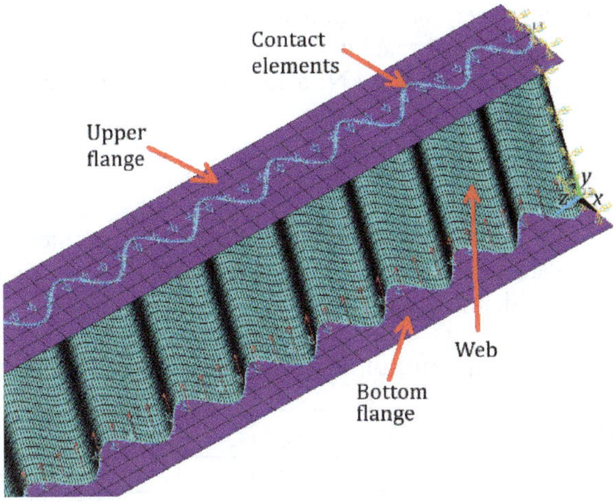

Figure 5: Finite element model of the spar beam and the boundary conditions.

4.2 Loading conditions

To realistically simulate the structural behaviour, the boundary conditions should be defined to reflect the physical loading conditions. The spars are rigidly fixed to the fuselage at one end and free at the other end. Accordingly, the displacements and rotations in all directions are restrained at the wing root in the model (Fig. 5), simulating the rigid joint between the wing and fuselage. The spars are supported by ribs and wing skin, which provide resistance to torsional loads and prevent local buckling in the flanges; therefore, no torsional load is applied to the spar model. To avoid twisting due to nonsymmetric web geometry under large transverse loads, z-direction movements at the central flange nodes are restrained (Fig. 5).

Lift loads are calculated based on the Hurkus Aircraft's specifications (maximum take-off weight (MTOW) of 3650 kg, +7 g/−3.5 g limits). With aircraft weight $W_A = g \times \text{MTOW}$, a critical positive limit ($g_L = +7\text{g}$), and a safety factor (SF) of 1.5 [5], the total critical lift load F_t is determined as 375 kN using:

$$F_t = SF \times g_L \times W_A. \quad (1)$$

Approximately 80% of this load is carried by the wings [10], and 75% of the wing lift is assumed to be borne by the main spar [5]. Thus, the lift load on one spar, F_s, is calculated as 112 kN. This load is applied upward on the bottom flange via a distributed formula [3]:

$$F(x_i) = \frac{F_0}{L}\sqrt{L^2 - x_i^2}, \quad (2)$$

where $F(x_i)$ is the force at a node with coordinate x_i, F_0 is the force applied to the node at the root, L is the spar length, and x_i is the coordinate of the nodes in the x-direction on which the force is applied. The value of F_0 is chosen so that the sum equals the full load of 200 kN.

The loads are applied on the central nodes of the bottom flange, considering that the wing skin uniformly bends and transmits a significant portion of the lift load to the stiffer central region of the spar.

Nonlinear buckling analyses are conducted using the arc-length method, which incrementally applies the load until reaching the full load or buckling. A load factor (λ) of 1.0 indicates full load resistance of 200 kN; a factor above 0.6 confirms that the spar can safely support the design load of 112 kN.

4.3 Material properties

IM7/8552 carbon-fibre-reinforced polymer, known for high stiffness-to-density ratios, is used for the spar. Its relevant material properties, as reported by Kaddour et al. [11], are given in Tables 1 and 2. The ply thickness was reported to be 0.125 mm [11].

Table 1: Elastic properties of unidirectional IM7/8552 laminae [11].

E_1(GPa)	E_2(GPa)	G_{12}(GPa)	ν_{12}	ν_{23}
165	9	5.6	0.34	0.499

Table 2: Strength properties unidirectional IM7/8552 laminae [11].

E_1(GPa)	E_2(GPa)	G_{12}(GPa)	ν_{12}	ν_{23}
165	9	5.6	0.34	0.499

4.4 Convergence study for mesh size and sub-step size

A mesh convergence study was conducted to balance numerical accuracy with computational efficiency. The analysis began with a coarse mesh, progressively refining it by halving the element size at each step. For each mesh configuration, a finite element (FE) buckling analysis of the spar was performed. Fig. 6 presents the resulting buckling loads for various element sizes used in the flanges and the web. Despite the notable differences in mesh resolution, the computed buckling loads exhibit minimal sensitivity to the element size. Based on these results, a maximum element size of 16.67 × 16.67 mm was selected for the flanges, while the web was discretized using an average element size of 2.55 × 10 mm. This configuration yields a total of 37,581 elements, providing a good compromise between solution fidelity and computational cost.

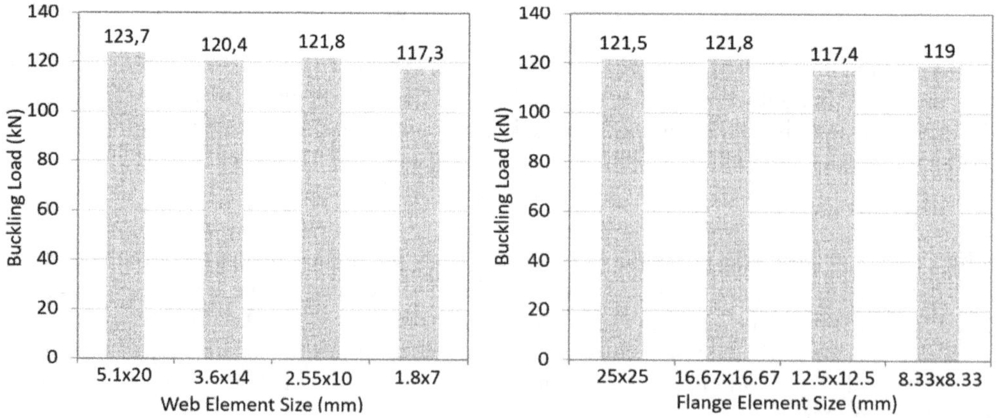

Figure 6: Sensitivity of the calculated buckling load on the finite element size.

In the finite element analysis, the load is applied incrementally using a specified number of sub-steps. A convergence study was conducted to determine the necessary number of sub-steps for accurate results. As Fig. 7 demonstrates 60 sub-steps for the full load of 200 kN, which realized as 37 sub-steps for the buckling load, yield sufficient accuracy while balancing computational time.

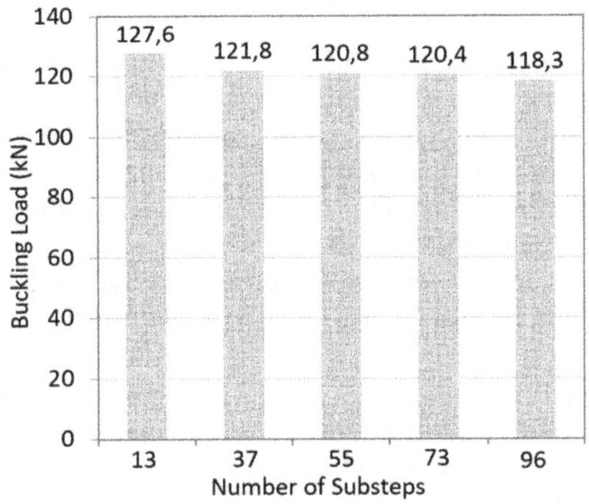

Figure 7: Sensitivity of the calculated buckling load on the sub-step number.

5 DESIGN OPTIMIZATION METHODOLOGY

5.1 Objective function and optimization variables

The goal is to minimize the mass of the corrugated web while ensuring its buckling load exceeds the lift load on the spar. The design variables are the four corrugation parameters – initial wavelength (w_1), wavelength increment rate (i), initial amplitude (a_1), and final amplitude (a_f) – which influence both mass and buckling resistance. Fibre orientation angles, θ_i, affect only the buckling load. Because of the different ways that the two types of optimization variables affect the performance of the spar beam, a two-level optimization approach is adopted.

In the first level, the objective is to minimize the mass of the web with the condition that the bucking load exceeds the lift load, F_s:

$$\text{Minimize } f_1(w_1, i, a_1, a_f) = \frac{m_C}{m_F} + \langle 1 - \frac{F_b}{F_s} \rangle, \tag{3}$$

where m_C and m_F are the masses of the corrugated and flat webs, respectively, and F_b is the buckling load. The second term is a penalty function accounting for constraint violations related to buckling failure. The bracket operator, < >, returns zero if its argument is negative; otherwise, the argument itself. If $F_b \geq F_s$, meaning the design has sufficient buckling strength, the penalty function has no effect.

In the second optimization level, the buckling load of the spar beam, F_b, is maximized, keeping the corrugation parameters fixed at their optimal values from the first level. Only the fibre orientations are used as optimization variables with stacking sequence $[\pm\theta_1/\pm\theta_2/\pm\theta_3]_s$. Then, the second-level optimization problem can be stated as

$$\text{Minimize } f_2(\theta_i) = -\frac{F_b}{F_s}. \tag{4}$$

By minimizing the negative of the F_b/F_s ratio, the algorithm effectively maximizes the buckling load, F_b.

5.2 Constraints on the optimization variables

Aside from the buckling strength constraint in the first optimization level, upper and lower bounds need to be imposed on the optimization variables. There may be geometric constraints; such as the initial amplitude, a_1, not exceeding half the flange width at the root, $b_r/2$. A manufacturing constraint is also imposed to prevent excessively dense corrugations with a small wavelength and large amplitude. Accordingly, the initial amplitude, a_1, must be smaller than a fraction of the initial wavelength, w_1, as $a_1 < w_1/3.5$.

The corrugated web geometry follows a decreasing trend toward the tip, as shown in Fig. 4. Therefore, the final amplitude, a_f, must be smaller than the initial amplitude, a_1, $(a_f < a_1)$.

Optimization could be performed without limits on the optimization variables other than behavioural constraints (e.g., buckling failure), or manufacturing constraints (e.g., small curvatures). However, searching across a large feasible domain, including the regions least expected to contain the global optimum configuration, would be computationally inefficient. On the other hand, if the search is conducted within a too-limited region, one may exclude the globally optimal or many near globally optimal configurations. In the present problem, a wide but reasonable range is chosen for the optimization variables with upper and lower bounds given in Table 3.

Table 3: Upper and lower bounds on the optimization variables.

Variable	Lower limit	Upper limit
Initial wavelength, w_1, (mm)	20	100
Wavelength increment rate, i	1.00	1.05
Initial amplitude, a_1, (mm)	1	35
Final amplitude, a_f, (mm)	0	25
Fiber angles, θ_i, (°)	15	75

5.3 Optimization algorithm

Structural optimization problems typically contain numerous locally optimal designs. To identify the best design, a global search algorithm is necessary. Accordingly, a modified simulated annealing (SA) algorithm, previously proposed in Akbulut and Sonmez [8] and Kaya and Sonmez [9] is employed. The modifications introduced to the algorithm improved its reliability and ease of applicability. Unlike the classical simulated annealing algorithm that requires manual tuning of parameters for the type of the optimization problem, the modified simulated annealing algorithm is designed to be self-tuning. Users need to specify only the upper and lower limits of the optimization variables, and the algorithm tunes itself to the specific problem at hand.

The key features and workflow of the algorithm are as follows:

1. Population-based search: Unlike classical SA, which perturbs a single solution, a population of $N = 9n$ configurations, where n is the number of the optimization variables, are maintained.
2. Initialization: N configurations are randomly generated within the prescribed upper and lower bounds. The objective functions of these configurations are evaluated and sorted in ascending order. The configuration in the first place is called the best, the last one is called the worst, and the one on the $7n$th place is called the worse configuration.
3. Generation of candidate configurations: At each iteration, a current configuration is randomly selected. For each optimization variable, x_i, a random value is generated in the neighbourhood of the selected current configuration using the following equation:

$$x_i' = x_i + r_i \Delta x_{\max}, \quad i = 1 \ldots n, \tag{5}$$

   where $\Delta x_{\max}$ is the current step size for variable x_i, r_i is a random number in $[-1, +1]$. If any x_i' violates its bound, it is regenerated until all the constraints are satisfied. In this way, a random candidate configuration with design parameters x_i' is obtained.
4. Objective function evaluation: The buckling load of the new configuration is calculated by carrying out a FE analysis, and its objective function value, f_n, is computed using eqn (3) or eqn (4).
5. Acceptance criterion: If f_n is less than the objective function value of the worse configuration, f_w, it is always accepted. Otherwise, its acceptability, A_n, is evaluated using:

$$A_n = \exp(-(f_n - f_w)/T_k), \tag{6}$$

   where T_k is the control parameter for the kth inner loop. If A_n is larger than a random number in $[0, +1]$, the new configuration is accepted.
6. Replacement: The first $7n$ configurations form a core set and the last $2n$ are designated replaceable. When a newly generated configuration is accepted, it replaces a randomly selected configuration among the last $2n$ configurations.
7. Initialization of the control parameter, T_k: Its initial value, T_0, is set so that the worst initial current configuration is accepted with 99% probability

$$T_0 = -(2f_{\text{worst}} - f_w)/\ln(0.99). \tag{7}$$

8. Initial step size: One-fourth of the difference between the upper and inner bounds.

$$\Delta x_{\max,0} = (x_u - x_l)/4. \tag{8}$$

9. Step-size reduction: In an inner loop, if the number of improvements in the first $7n$ configurations is less than 10% of the number of trials, L^k, the maximum variations in the variables, $\Delta x_{\max}$ is reduced by 5%.
10. Inner-Loop Length: The number of trials at a certain inner loop is:

$$L^k = \text{round}\{3n[1 + 2(\Delta x_{\max}/\Delta x_{\max.0})]\}. \tag{9}$$

    That means it starts at $9n$, and decreases towards $3n$ as the step size, $\Delta x_{\max}$, shrinks.
11. Adaptive cooling: After each inner loop, the control parameter, T_k, is reduced by multiplying it by the reduction factor, α_k:

$$T_{k+1} = \alpha_k T_k. \tag{10}$$

Its value is calculated using the following equation:

$$\alpha_k = \begin{cases} 0.8 & \text{if } L_{\mathrm{acc}}^k / L^k > \Delta x_{\max} / \Delta x_{\max,0} \\ 0.999 & \text{else,} \end{cases} \tag{11}$$

where L_{acc}^k is the number of accepted trials in the inner loop, L_{acc}^k / L^k is the acceptance ratio, $\Delta x_{\max}$ is the current step size, $\Delta x_{\max,0}$ is the initial step size.

12. Termination: Iterations continue until the difference between the objective function values of the best and worst current configurations falls below a small threshold, indicating convergence.

Because all control parameters – including initial temperature, cooling schedule, step size, and inner-loop length – are derived automatically from the variable bounds and observed acceptance rates, this modified SA requires no manual tuning beyond specifying the search domain. The result is a robust, efficient optimizer that balances global exploration and fine-scale refinement with minimal user intervention.

6 RESULTS AND DISCUSSION

A code was developed using ANSYS Parametric Design Language (APDL) to implement the modified simulated annealing algorithm and carry out FE analyses. Using the code, trial configurations are generated through iterations and the buckling load of each one is calculated until convergence to the optimal one is achieved. Comparisons are made between the spar beams with the optimal corrugated web and flat web regarding buckling strength and mass.

6.1 Spar with the optimal corrugated web

In the first optimization step, the mass of the web is minimized by varying only the parameters defining its corrugated geometry while ensuring adequate buckling strength. The layup sequence of the web is taken as $[\pm 45]_{3s}$ in the first step. Starting from initial configurations randomly generated within the bounds given in Table 3, the modified simulated annealing algorithm [8], [9] iterated until convergence. The optimal corrugation parameters (Table 4) yield a buckling load of 120.7 kN, exceeding the required design load, $F_s = 112$ kN.

Table 4: Optimal corrugation parameters from first-level optimization.

First-level optimization		**Second-level optimization**	
Variable	**Lower limit**	**Variable**	**Lower limit**
Initial wavelength, w_1, (mm)	65.3	θ_1, (deg)	26
Wavelength increment rate, i	1.01	θ_2, (deg)	57
Initial amplitude, a_1, (mm)	3.73	θ_3, (deg)	61
Final amplitude, a_f, (mm)	0.64		
Buckling load (kN)	120.7	Buckling load (kN)	149

In the second optimization step, the buckling load is maximized by varying only the fibre orientation angles of the web, while the corrugation geometry obtained in the first step is fixed. The resulting optimal fibre angles are provided in Table 4 for web layup $[\pm\theta_1/\pm\theta_2/\pm\theta_3]_s$. The second-level optimization increases the buckling strength to 149 kN, significantly higher than the design load. Fig. 8 shows the buckling mode. Over half the corrugations buckle concurrently, indicating a structurally efficient design where weak spots have been eliminated, leading to a uniform structural response. Near the tip, no signs of buckling are observed, suggesting the possibility of removing corrugations or even thinning the web in that region using ply drops.

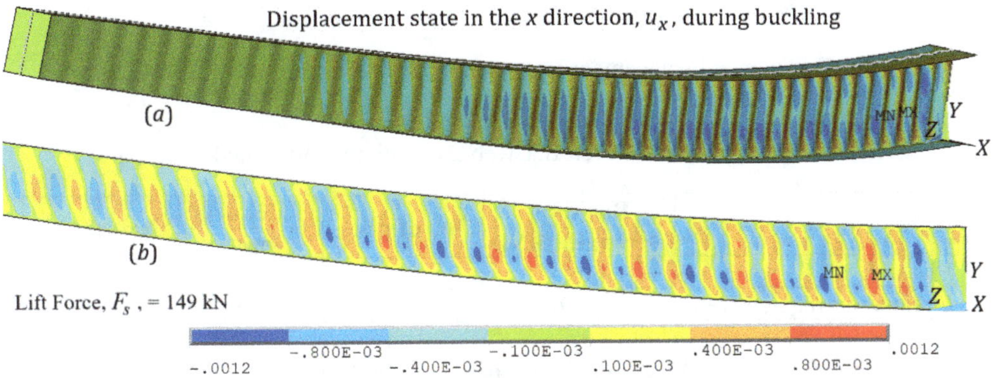

Figure 8: Transverse displacement u_x (m) of the optimized spar during buckling under a 149 kN lift load on the spar, F_s. (a) Isometric view; and (b) Close-up right view.

6.2 Spar with a flat web

Spars with flat webs exhibit much lower buckling resistance and thus require substantially thicker laminates to match corrugated designs. To compare the mass of the optimal spar design with that of the conventional spar with a flat web having about the same buckling resistance, finite element analyses of spars with a flat web having various thicknesses are carried out. As mentioned before, a perfectly flat web does not exhibit buckling behaviour in finite element simulations. A geometric feature triggering the expected buckling response is needed. For this purpose, a corrugation with 2.0 m wavelength and 0.1 mm amplitude is introduced to the web. This geometric irregularity can be considered within the range of manufacturing errors.

A 36-ply flat web, three times thicker than the optimized web, yields a buckling load of 146.4 kN, comparable to the optimized corrugated design. Fig. 9 displays its buckling mode, showing localized failure near the root. Table 5 compares the spars with the optimal corrugated web and flat web. The optimal corrugation significantly improves the buckling strength so that one may safely use thin composite laminates without the risk of buckling.

7 CONCLUSIONS

The study presents a design optimization methodology for composite I-beam wing spars featuring a variable, wavy corrugated web to achieve minimum mass. The web is modelled as a wave whose wavelength and amplitude vary along its length. The lift, the most dominant load carried by spars, is applied directly to the spar model. Since shear buckling is the primary failure mode for thin webs, a nonlinear buckling analysis is performed using an incremental

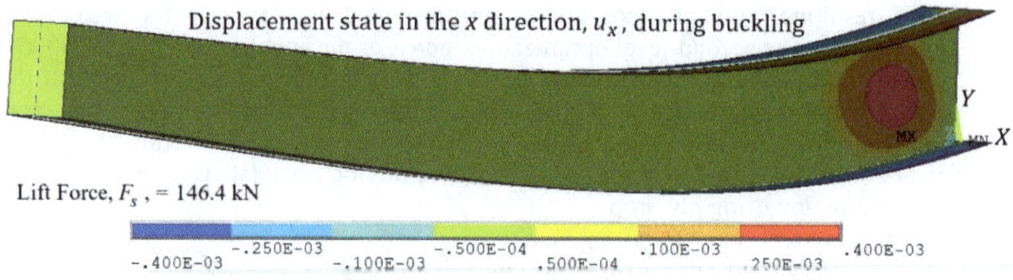

Figure 9: Transverse displacement u_x (m) of a flat-web spar under a 146.4 kN load.

Table 5: Comparison between flat-web and optimal corrugated-web spars.

	Buckling strength (kN)	**No. of plies**	**Thickness (mm)**	**Mass (kg)**
Flat web	146.4	36	4.5	4.291
Optimal corrugated web	149	12	1.5	1.444

loading strategy via the arc-length method. Mesh density, sub-step numbers, and multiplier values for the arc-length analysis were all refined through convergence studies, yielding results within acceptable accuracy limits.

A two-step optimization approach is employed. In the first stage, the mass is minimized by optimizing the corrugation parameters. In the second stage, with the corrugation parameters fixed at their optimal values, the fibre orientation angles are optimized to maximize the buckling load. The optimal corrugated web spar exhibits a significantly higher shear buckling capacity; notably, a flat-web spar achieving comparable buckling strength is approximately three times heavier.

It is important to note that the present work focuses on component-based optimization. In industrial applications, the optimized spar should be integrated into a complete wing model, its mechanical response should be assessed under realistic loading conditions, and then the optimization process should be repeated using the load distributions obtained from the full-wing analysis.

ACKNOWLEDGEMENT

This study is supported by Bogazici University Research Fund with grant number 18A06P1.

REFERENCES

[1] Kumaraswamy, T., Analysis and fabrication of composite wing spar. *Int. J. Mag. Eng. Technol. Manag. Res.*, **3**(9), pp. 648–661, 2016.

[2] James, R., What are the parts of an airplane wing? Pilot Teacher. https://pilotteacher.com/what-are-the-parts-of-an-airplane-wing/. Accessed on: 1 Apr. 2025.

[3] Visnjic, G., Nožak, D., Kosel, F. & Kosel, T., Reducing shear-lag in thin-walled composite I-beam wing spars. *Aircr. Eng. Aerosp. Technol.*, **86**(2), pp. 89–98, 2014.

[4] Elangovan, S., Sureshkumar, C. & Divyabarathi, P., Design and analysis of aircraft wing spar with different materials using ansys. *Int. J. Mech. Prod. Eng. Res. Dev.*, **9**(3), pp. 613–618, 2019.

[5] Ajith, S., Paramasivam, R. & Vidhya, K., Study of optimal design of spar beam for the wing of an aircraft. *Int. J. Eng. Dev. Res.*, **5**(3), pp. 179–193, 2017.

[6] Bhachu, K.S., Haftka, R.T., Waycaster, G., Kim, N.H. & Hurst, C., Probabilistic manufacturing tolerance optimization of damage-tolerant aircraft structures using measured data. *J. Aircr.*, **52**(5), pp. 1412–1421, 2015.

[7] Best, P.E., Analysis and comparison of I-beam and solid wing spars. *Sport Aviat.*, pp. 38–42, 1961.

[8] Akbulut, M. & Sonmez, F.O., Design optimization of laminated composites using a new variant of simulated annealing. *Comput Struct.*, **89**(17–18), pp. 1712–1724, 2011.

[9] Kaya, S. & Sonmez, F.O., Optimum local reinforcements for maximum strength of composite plates with a cutout. Submitted to *Struct. Multidiscip. Optim.*

[10] Shabeer, K.P. & Murtaza, M.A., Optimization of aircraft wing with composite material. *Int. J. Innov. Res. Sci. Eng. Technol.*, **2**, pp. 2471–2477, 2013.

[11] Kaddour, A.S., Hinton, M.J., Smith, P.A. & Li, S., Mechanical properties and details of composite laminates for the test cases used in the third world-wide failure exercise. *J. Compos. Mater.*, **47**(20–21), pp. 2427–2442, 2013.

Author index

www.ingramcontent.com/pod-product-compliance
Lightning Source LLC
LaVergne TN
LVHW062237200926
848484LV00006B/1562

* 9 7 8 1 7 8 4 6 6 4 9 7 8 *